MANUFACTURING TECHNOLOGY RESEARCH

ADVANCED MANUFACTURING IN AMERICA

OPPORTUNITIES AND STRATEGIES

Manufacturing Technology Research

Additional books in this series can be found on Nova's website under the Series tab.

Additional E-books in this series can be found on Nova's website under the E-book tab.

Economic Issues, Problems and Perspectives

Additional books in this series can be found on Nova's website under the Series tab.

Additional E-books in this series can be found on Nova's website under the E-book tab.

MANUFACTURING TECHNOLOGY RESEARCH

ADVANCED MANUFACTURING IN AMERICA

OPPORTUNITIES AND STRATEGIES

BRANDON CULLERNE
AND
MICHAEL LASTON
EDITORS

nova publishers
New York

For permission to use material from this book please contact us:
Telephone 631-231-7269; Fax 631-231-8175
Web Site: http://www.novapublishers.com

Library of Congress Cataloging-in-Publication Data

ISBN: 978-1-62257-932-7

Published by Nova Science Publishers, Inc. † New York

Contents

PREFACE

This book is a study on the developments of a strategic plan to guide Federal programs and activities in support of advanced manufacturing research and development. Advanced manufacturing is a matter of fundamental importance to the economic strength and national security of the United States. Analysis of patterns and trends in U.S. advanced manufacturing reveals both opportunities for Federal policy to accelerate the development of this vital sector and challenges to its continuing health. The acceleration of innovation for advanced manufacturing requires bridging a number of gaps in the present U.S. innovation system, particularly the gap between research and development (R&D) activities and the deployment of technological innovations in domestic production of goods. The strategic plan discussed in this book lays out a robust innovation policy that would help to close these gaps and address the full lifecycle of technology.

Chapter 1 – This report responds to Section 102 of the America COMPETES Reauthorization Act of 2010, which directs the Committee on Technology of the National Science and Technology Council (NSTC) to develop a strategic plan to guide Federal programs and activities in support of advanced manufacturing research and development. Advanced manufacturing is a matter of fundamental importance to the economic strength and national security of the United States. The authors' analysis of patterns and trends in U.S. advanced manufacturing reveals both opportunities for Federal policy to accelerate the development of this vital sector and challenges to its continuing health.

The acceleration of innovation for advanced manufacturing requires bridging a number of gaps in the present U.S. innovation system, particularly the gap between research and development (R&D) activities and the deployment of technological innovations in domestic production of goods. This strategic plan lays out a robust innovation policy that would help to close these gaps and address the full lifecycle of technology. It also incorporates intensive engagement among industry, labor, academia, and government at the national, state, and regional levels. Partnerships among diverse stakeholders, varying by location and objective, are a keystone of the strategy.

The strategy seeks to achieve five objectives. These objectives are interconnected; progress on any one will make progress on the others easier. A large number of Federal agencies, coordinated through the NSTC, have important roles to play in the implementation of the strategy.

Objective 1: *Accelerate investment in advanced manufacturing technology, especially by small and medium-sized manufacturing enterprises, by fostering more effective use of Federal capabilities and facilities, including early procurement by Federal agencies of cutting-edge products.*

Objective 2: *Expand the number of workers who have the skills needed by a growing advanced manufacturing sector and make the education and training system more responsive to the demand for skills.*

Objective 3: *Create and support national and regional public-private, government-industry-academic partnerships to accelerate investment in and deployment of advanced manufacturing technologies.*

Objective 4: *Optimize the Federal government's advanced manufacturing investment by taking a portfolio perspective across agencies and adjusting accordingly.*

Objective 5: *Increase total U.S. public and private investments in advanced manufacturing research and development (R&D).*

Chapter 2 – Advanced manufacturing is not limited to emerging technologies; rather, it is composed of efficient, productive, highly integrated, tightly controlled processes across a spectrum of globally competitive U.S. manufacturers and suppliers. For advanced manufacturing to accelerate and thrive in the United States, it will require the active participation of communities, educators, workers, and businesses, as well as Federal, State, and local governments.

The Advanced Manufacturing Partnership (AMP) Steering Committee proposes that the Nation establish a national advanced manufacturing strategy. This strategy will servc as a national framework that, when implemented by states and local communities, will bring about a sustainable resurgence in advanced manufacturing in the United States.

The AMP Steering Committee developed a set of 16 recommendations around three pillars:

- Enabling innovation
- Securing the talent pipeline
- Improving the business climate

These recommendations are aimed at reinventing manufacturing in a way that ensures U.S. competitiveness, feeds into the Nation's innovation economy, and invigorates the domestic manufacturing base. The objective is to position the Nation to lead the world in new disruptive advanced manufacturing technologies that are changing the face of manufacturing.

The AMP Steering Committee believes that a number of important steps taken now will be critical to strengthen the Nation's innovation system for advanced manufacturing. While some of the largest U.S. firms have the depth and resources to be ready for this challenge, a significant number of small and medium-sized U.S. firms operate largely outside the present innovation system. The United States will only lead in advanced manufacturing if all companies are able to participate in the transformations made possible through innovations in manufacturing.

The AMP Steering Committee proposes 16 recommendations that will set the stage for advanced manufacturing to thrive in the United States

Chapter 3 – The Technology Development Workstream had two broad objectives. The first was to determine a sustaining mechanism to be used for identifying and developing key manufacturing technologies, and the second was to set forth a set of top technology areas that would ensure U.S. manufacturing competitiveness.

To meet the objectives, the mechanisms used to gather data were mainly the following:

- Surveys to elicit opinions and feedback from industry and university participants regarding important industries for the future, and key technologies that will be needed, as well as private-public partnership best practice examples and models.
- Workshops with industry and university participants to identify the top technologies required for manufacturing competitiveness.
- Desk research of comparable mechanisms used in other countries and regions to identify and nurture technologies.
- Whitepaper solicitations from select experts.

The research and innovation ecosystem of a nation is highly dependent on the presence of a manufacturing base that provides constant feedback in terms of problems and challenges to be solved. Also, product innovation cannot exist without intimate knowledge and control over the manufacturing process. In other words, design of the product also inherently involves design of the manufacturing process by which the product will be made. Historically, the United States has had a vibrant manufacturing base and active programs in both basic and applied research. The distinguishing feature of U.S. research activity has been the sheer breadth and vitality at various levels.

A review of mechanisms used in other countries indicate a use of more hierarchical, systematic planning systems that are explicitly aligned to national strategies– often necessitated by their lack of resources, as opposed to in the United States, where much more bottom-up innovation flourishes and much larger bets are made with respect to research. But in 2011, the world shifted. For the first time in history, total R&D spending in Asia Pacific exceeded total R&D spending in the U.S., with the largest increases occurring in China from government sources. This trend is forecasted to continue at even a higher rate. The use of a structured planning process at a national level has the advantages of both aligning and allocating national resources more efficiently into U.S. efforts to revitalize planning, as well as creating a platform to better address competing national strategies from other countries (that are often government led and therefore difficult for U.S. industry participants to counter by themselves). Therefore, the authors have incorporated this kind of a hierarchical planning process into the authors' recommendations.

Chapter 4 – In June 2011, the President established the Advanced Manufacturing Partnership (AMP), which is led by a Steering Committee that operates within the framework of the President's Council of Advisors on Science and Technology. In July 2012, the AMP Steering Committee delivered its report to PCAST, entitled *Capturing Domestic Competitive Advantage in Advanced Manufacturing*. PCAST adopted this report and submitted it to the President. The Steering Committee's report draws on preliminary reports prepared by several "workstreams." These workstream reports have been made available as on-line annexes to the Steering Committee report.

Chapter 5 – For the United States to remain a competitive force on the world stage, talented employees who have a high level of technical skill are needed to revitalize, sustain, and improve U.S. manufacturing. Unfortunately, the image of manufacturing and the public perception that it can provide long-term, desirable careers have been tarnished.

This negative image is driving the most talented, technically skilled students away from manufacturing to other career paths and creating a deficiency in the quantity and quality of the current and future workforce. To attract a robust and highly skilled workforce, the image of manufacturing must change from offering low job security and dull, dirty, and dangerous work to being exciting, engaging, essential, and environmentally sustainable. The same cohesive message from the government, educational institutions, and private industry is needed to change this perception. An equally important need is for some modification to traditional teaching methods used to train the manufacturing workforce at all levels of education. Success in advanced manufacturing and entrepreneurship will require a workforce with fundamental science, technology, engineering, and math (STEM) skills and broad problem-solving skills, decision-making skills, and people skills that do not emerge from a conventional K–12 education. The authors encourage adoption of Project-Based Learning (PBL) methods in upper K–12 and in community college programs in manufacturing, with some projects selected for their relevance to manufacturing-relevant skills, such as supply-chain management, design for manufacturability, estimation of tolerances and requirements, economics, and team-management. To stimulate these new educational initiatives, educational partnerships between industry, academia, and local and regional governments must be established. A successful rebirth of manufacturing will be ensured only by addressing the underlying structural challenges.

Chapter 6 – The United States is at risk of losing leadership in manufacturing. This is true not only in low-technology industries and products, but also with respect to the authors' ability to manufacture the high-technology products that are invented and innovated in this country. The authors do not believe that it is the role of government to formulate a national industrial policy of direct investment in or subsidies to specific firms. However, the authors believe strongly that the United States requires a coherent national policy framework oriented toward creating a favorable business climate for manufacturing that spurs investments and fosters partnerships between government, academia, and industry. The manufacturing sector must have a competitive domestic environment, which includes a robust talent pipeline, physical capital, and intellectual capital.

While these capabilities are important to the overall health of the U.S. economy, they are particularly necessary for the advanced manufacturing sector, which faces intense global competition. Every major economic competitor is taking steps to attract investment in advanced manufacturing, recognizing the impacts of that investment on the broader economy.

Given the strong link between innovation and advanced manufacturing, the Partnership has put forward a package of synergistic recommendations to encourage greater U.S.-based innovation, research and development (R&D), and investment in advanced manufacturing. For the United States to continue to be an attractive location for businesses, it is important to build a policy framework that spurs investments and fosters partnerships between government, academia, and industry. The foundation of that framework is constructed through targeted policies in four areas that have a significant impact on advanced manufacturing:

- Tax policy,
- Smarter regulations,
- Trade policy, and
- Energy policy.

The Advanced Manufacturing Partnership (AMP) Steering Committee (SC) has developed additional specific recommendations in three areas to improve the climate for robust industry/ university collaboration in research and commercialization, with the aim of reinvigorating what has been an historic strength of the United States:

- Remove barriers in tax-exempt buildings at universities to enable expanded university-industry collaborations,
- Expand capital resources for emerging advanced manufacturing enterprises, and
- Coordinate and expand existing programs that enable university researchers to interact directly with manufacturers.

Chapter 7 – Between September 2011 and March 2012, the Outreach Workstream held a series of meetings with stakeholders who have an interest in advanced manufacturing policy. Stakeholders included Congressional staff, trade associations, scientific associations, and think tanks. These meetings had three purposes: (1) to educate stakeholders about the Advanced Manufacturing Partnership (AMP), (2) to solicit feedback about issues on which the AMP Steering Committee (SC) should focus, and (3) to learn about what their respective organizations are doing that could inform AMP's work and be a resource for the Advanced Manufacturin gNational Program Office (NPO).

Chapter 8 – The Advanced Manufacturing Partnership (AMP) is a national effort bringing together the Federal government, industry, universities, and other stakeholders to identify and invest in emerging technologies with the potential to create high-quality domestic manufacturing jobs and enhance the global competitiveness of the United States. AMP responds to recommendations made by the President's Council of Advisors on Science and Technology (PCAST) in its June 2011 report, entitled "Ensuring Leadership in Advanced Manufacturing." AMP is guided by a Steering Committee, which operates within the framework of PCAST and is comprised of leading experts from industry and academia. The AMP Steering Committee (AMP SC) is organized by four workstreams: Technology Development; Policy; Education and Workforce Development; and Shared Facilities and Infrastructure. In the fall of 2011, the AMP SC hosted four regional outreach meetings around the country and over 1000 members of the public representing diverse stakeholder perspectives participated:

- Atlanta, GA, October 14, 2011 (hosted by the Georgia Institute of Technology)
- Cambridge, MA, November 28, 2011 (hosted by Massachusetts Institute of Technology)
- Berkeley, CA, December 5, 2011 (hosted by University of California - Berkeley and Stanford University)
- Ann Arbor, MI, December 12, 2011 (hosted by the University of Michigan)

The regional meetings had two purposes. The first was to share AMP's approach and activities with industry, university, government, and other stakeholders. The second was to gather the participants' ideas about opportunities for investments and actions that have the potential to transform manufacturing in the United States. Participants at each of the four regional meetings heard from government and industry panelists, who discussed opportunities and challenges that their communities face in the advanced manufacturing domain. These panel discussions were followed by breakout sessions focused on the themes of the four workstreams. The main points of these breakout sessions are synthesized below. Specific details on each meeting can be found in the Appendix.

Chapter 9 – Manufacturing plays an important role in the nation's economy, employment, and national defense. Accordingly, Congress has maintained a strong interest in the health of the U.S. manufacturing sector. Some analysts have expressed concerns about a decades-long decline in manufacturing employment punctuated by a steep drop from 2001 to 2010, as well as about the offshore outsourcing of production and related functions, such as research and development, by U.S. manufacturers. Others see the U.S. manufacturing sector as vibrant and healthy as evidenced by growth in output and productivity.

The Obama Administration has undertaken a number of initiatives intended to support U.S. manufacturing, including establishment of the Advanced Manufacturing Partnership, Advanced Manufacturing National Program Office (AMNPO), Advanced Manufacturing Technology Consortia program, National Robotics Initiative, and Materials Genome Initiative. In his FY2013 budget, President Obama proposed the creation of a National Network for Manufacturing Innovation (NNMI) to help accelerate innovation by investing in industrially relevant manufacturing technologies with broad applications, and to support manufacturing technology commercialization by bridging the gap between the laboratory and the market. The NNMI proposal calls for the establishment of up to 15 Institutes for Manufacturing Innovation (IMI) funded through a one-time infusion of $1 billion in mandatory funding to the Department of Commerce's National Institute for Standards and Technology (NIST) and carried out over a period of 10 years. Each IMI would be comprised of stakeholders from industry (including large companies and small- and medium-sized manufacturing enterprises), academia, federal agencies, and state government entities. According to the proposal, each IMI is to be competitively selected, serve as a regional hub for manufacturing innovation (as well as part of the national network), and have a unique focus area (e.g., an advanced material, manufacturing process, enabling technology, or industry sector). The NNMI would be managed collaboratively by NIST, the Department of Defense, Department of Energy, National Science Foundation, and other agencies. An Administration official has testified that the NNMI program has not yet begun and that the establishment of the NNMI will require congressional authorization and funding. No legislation has been introduced to establish the NNMI. However, the President announced his intention to establish a pilot institute using current year funding; in this regard, the Department of Defense issued a Broad Agency Announcement (BAA) in May 2012 to award up to $30 million for an institute focused on additive manufacturing. The pilot is described as "the first institute to be launched within the National Network for Manufacturing Innovation" and "a proof-of-concept for the potential subsequent institutes." The BAA limits the types of organizations that can compete for the award of this institute to universities and non-profit organizations. Universities must incorporate in their proposals a partnership with a cluster of manufacturing firms and associated institutes; non-profit industry consortia must incorporate a partnership with universities for applied research. The AMNPO has published a request for information seeking public comment on various aspects of the NNMI to aid in the development of the institutes and the network. Topics included in the RFI include technology focus areas, structure, governance, sustainability, and education and workforce development.

In: Advanced Manufacturing in America
Editors: Brandon Cullerne and Michael Laston
ISBN: 978-1-62257-932-7

Chapter 1

A National Strategic Plan for Advanced Manufacturing*

National Science and Technology Council

Executive Summary

This report responds to Section 102 of the America COMPETES Reauthorization Act of 2010, which directs the Committee on Technology of the National Science and Technology Council (NSTC) to develop a strategic plan to guide Federal programs and activities in support of advanced manufacturing research and development. Advanced manufacturing is a matter of fundamental importance to the economic strength and national security of the United States. Our analysis of patterns and trends in U.S. advanced manufacturing reveals both opportunities for Federal policy to accelerate the development of this vital sector and challenges to its continuing health.

The acceleration of innovation for advanced manufacturing requires bridging a number of gaps in the present U.S. innovation system, particularly the gap between research and development (R&D) activities and the deployment of technological innovations in domestic production of goods. This strategic plan lays out a robust innovation policy that would help to close these gaps and address the full lifecycle of technology. It also incorporates intensive engagement among industry, labor, academia, and government at the national, state, and regional levels. Partnerships among diverse stakeholders, varying by location and objective, are a keystone of the strategy.

The strategy seeks to achieve five objectives. These objectives are interconnected; progress on any one will make progress on the others easier. A large number of Federal agencies, coordinated through the NSTC, have important roles to play in the implementation of the strategy.

* This is an edited, reformatted and augmented version of the National Science and Technology Council, dated February 2012.

Objective 1: *Accelerate investment in advanced manufacturing technology, especially by small and medium-sized manufacturing enterprises, by fostering more effective use of Federal capabilities and facilities, including early procurement by Federal agencies of cutting-edge products.*

Objective 2: *Expand the number of workers who have the skills needed by a growing advanced manufacturing sector and make the education and training system more responsive to the demand for skills.*

Objective 3: *Create and support national and regional public-private, government-industry-academic partnerships to accelerate investment in and deployment of advanced manufacturing technologies.*

Objective 4: *Optimize the Federal government's advanced manufacturing investment by taking a portfolio perspective across agencies and adjusting accordingly.*

Objective 5: *Increase total U.S. public and private investments in advanced manufacturing research and development (R&D).*

1. INTRODUCTION

Section 102 of the America COMPETES Reauthorization Act of 2010 directs the Committee on Technology of the National Science and Technology Council (NSTC) to develop a strategic plan to guide Federal programs and activities in support of advanced manufacturing research and development (see Appendix A). This report responds to that provision. It was prepared by the Interagency Working Group on Advanced Manufacturing (IAM) of the NSTC, which included representatives of Federal agencies with a vital interest in sustaining and strengthening the Nation's advanced manufacturing sector.

Advanced manufacturing is a family of activities that (a) depend on the use and coordination of information, automation, computation, software, sensing, and networking, and/or (b) make use of cutting edge materials and emerging capabilities enabled by the physical and biological sciences, for example nanotechnology, chemistry, and biology. It involves both new ways to manufacture existing products, and the manufacture of new products emerging from new advanced technologies.[1]

Advanced manufacturing is a matter of fundamental importance to the economic strength and national security of the United States. Advanced manufacturing provides high-quality jobs. It is an important source of exports. It is a key source of technological innovation. It provides essential goods and equipment for the military, the intelligence community, and homeland security agencies. These impacts justify Congressional and executive branch attention to Federal policies that affect advanced manufacturing.

Our analysis of patterns and trends in U.S. advanced manufacturing reveals both opportunities for Federal policy to accelerate the development of this vital sector and challenges to its continuing health. In particular, a gap exists between research and development (R&D) activities and the deployment of technological innovations in domestic production of goods.[2] This gap has contributed to the erosion of key indicators, such as the balance of trade in advanced technology products as measured by the U.S. Census Bureau (see Figure 1).[3] The United States ran a trade surplus in this category throughout the 1990s, but by 2010, that surplus had become an $81 billion deficit.

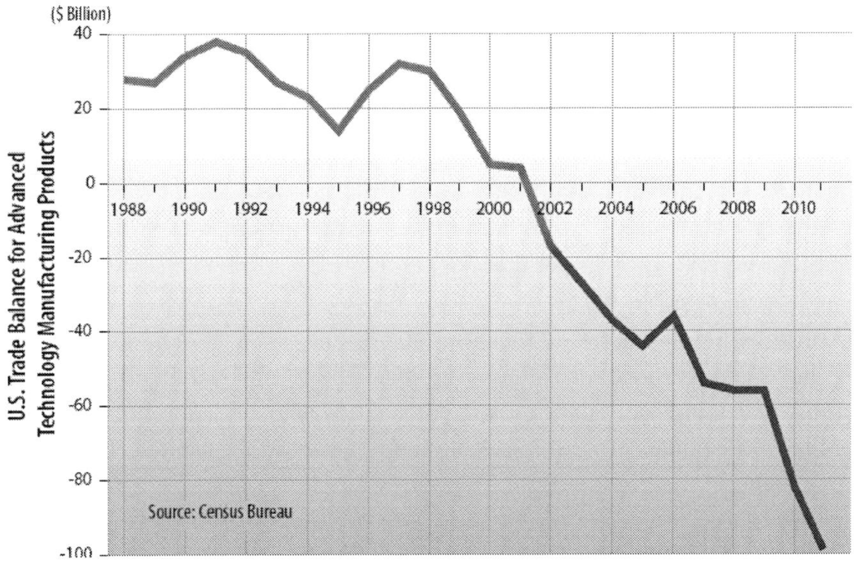

Figure 1. U.S. Trade Balance for Advanced Technology Products.

This strategic plan lays out a robust innovation policy that would reduce the gap between R&D and deployment of advanced manufacturing innovations. Specifically, this policy would address the full lifecycle of technology in order to (1) provide a fertile innovation environment for advanced manufacturing, (2) enable vigorous domestic development of transformative manufacturing technologies, (3) promote coordinated public and private investment in precompetitive advanced manufacturing technology infrastructure, and (4) facilitate rapid scale-up and market penetration of advanced manufacturing technologies.[4]

Federal policies and programs alone cannot address the Nation's challenges in advanced manufacturing. The strategic plan therefore incorporates intensive engagement among industrial, labor, academic, and government stakeholders at the national, state and regional levels.

The next section describes the challenges and opportunities for advanced manufacturing in more detail. Section 3 articulates the broad approach of the strategic plan and its five objectives. The remaining sections elaborate on how each of these objectives might be achieved.

2. ADVANCED MANUFACTURING: PATTERNS AND TRENDS

U.S. manufacturers produced about $1.7 trillion of goods in 2010, about 11.7% of the U.S. gross domestic product (GDP).[5] They employed 11.5 million Americans in jobs that paid on average about 21% more than average hourly compensation in private-sector service industries.[6] Manufacturing has a larger multiplier effect than any other major economic activity -- a dollar spent in manufacturing drives an additional $1.35 in economic activity.[7] Manufacturing is also the largest contributor to U.S. exports. In 2010, the United States exported over $1.1 trillion of manufactured goods, which accounted for 86% of all U.S. goods exports and 60% of U.S. total exports.[8]

Manufacturing provides many of the jobs and drives many of the businesses of today. Yet its role in providing the jobs and driving the businesses of tomorrow is even more important. The manufacturing sector accounts for about 72% of all private-sector R&D spending and employs about 60% of U.S. industry's R&D workforce.[9] As a result, the manufacturing sector develops and produces many of the technologies that advance the competitiveness and growth of the entire economy, including the much larger service sector. Technology-based improvements to productivity made possible by the manufacturing sector consistently generate job growth over time across the economy.[10]

Advanced manufacturing is emerging as an especially potent driver of future economic growth. A distinguishing feature of advanced manufacturing is its continual improvement in processes and rapid introduction of new products. It is this paradigm-shifting aspect of advanced manufacturing that has the most potential to spin off entirely new industries and lead to production methods that are most likely to "stick" in the United States because they are hard to imitate.

Global Trends in Advanced Manufacturing

Current global trends in R&D, innovation, and trade raise concerns about U.S. competitiveness in advanced manufacturing. In 2009, the United States ranked eighth among industrialized nations for R&D intensity (defined as national R&D as a share of GDP), according to the Organisation for Economic Co-operation and Development (OECD).[11] A 2011 report by the Information Technology and Innovation Foundation ranked the United States fourth out of 44 industrialized countries and regions in global innovative-based competitiveness, but second-to-last in progress toward increasing innovation-based competitiveness and capacity since 2000.[12]

As we noted in the Introduction, the Nation's trade balance for advanced technology products has deteriorated precipitously over the past decade, despite an offsetting 34% decline in the major-currency foreign exchange value of the U.S. dollar.[13] Currently, Germany, Korea, and Japan each have more R&D-intensive manufacturing sectors than the United States (see Figure 2);[14] moreover, they each have positive trade balances in goods.

Global competition in advanced manufacturing is growing more intense as technology lifecycles are accelerated. A lifecycle starts with the basic concept for a technology. For many technologies, scientific knowledge created through basic research provides key insights that enable the basic concept. Then the concept is validated, often through applied research. Next, development efforts mature the technology into a prototype of a commercial product. Finally, commercialization and scale-up activities convert the prototype into a commercially viable product and increase the scale of production to an economically viable level, respectively. Companies typically recapture the costs of technology investment through profits in the scale-up phase.

Acceleration of the lifecycle increases the importance of gaining market share in the commercialization phase, so that domestic manufacturers can seize the huge opportunities associated with the scale-up phase. In addition to achieving economies of scale, scale-up today requires achieving economies of scope by rapidly producing alternative versions of the same core product to satisfy diverse global customers.

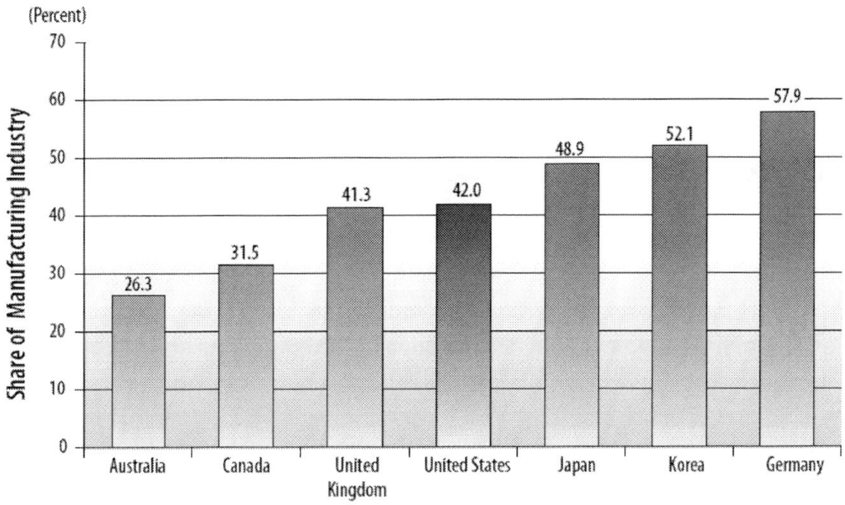

Source: OECD STAN Indicators 2009, "Value-added shares relative to manufacturing," Percent of manufacturing sector with 3% or greater R&D intensity, stats.oecd.org/index.aspx?r=228903

Figure 2. Share of Manufacturing Value-Added by Research-Intensive Manufacturing Sectors (R&D > 3% of Sales).

Manufacturing capability gaps in the United States have led to the loss of substantial economic benefits. American researchers invented and commercialized industrial robots, for examples, with the first installation in a General Motors plant in 1961, but now the vast bulk of industrial robot production is done in Asia and Europe. The same pattern holds in energy storage and power generation, and in many other areas of technology. U.S.-based facilities no longer produce electronic displays for computer monitors, televisions, or handheld devices, such as the Kindle e-reader.[15]

The loss of such production capabilities affects U.S. national security as well as the national economy. The Defense Production Act Committee (DPAC) (see Appendix B) has identified a number of vital government needs that the United States is currently unable to meet via secure and reliable domestic production. DPAC analysis also revealed several key technologies for which an impending lack of U.S. production will likely create national security vulnerabilities. A sampling of specific vulnerabilities include aircraft landing gear; railcar components; large rotor disks for turbines; rocket engine parts; missile launch systems; unmanned aerial and ground vehicles; nuclear power components; aircraft fuselages; orbital vehicles; network routing and switching; optical data transport; advanced power electronics; low cost composites; and transmission conductors.

3. Principles and Objectives of the National Strategy

Challenges to the competitiveness of U.S. advanced manufacturing have the potential to undermine the Nation's ability to create jobs, invent new industries, and protect itself from security threats in the 21st century. The United States should respond to these challenges with an *innovation policy for advanced manufacturing* that addresses (1) the complexity of public and private elements of modern manufacturing technologies, and (2) the ways in which these elements change over the technology lifecycle. By doing so, the Nation can, as Secretary of Commerce John Bryson put it recently, "build it here, sell it everywhere."[16]

Innovation Policy for Advanced Manufacturing

An innovation policy for advanced manufacturing must be capable of responding to a range of market failures. Markets sometimes fail to provide adequate incentives to manufacturers to make specific investments that would benefit the economy over the long-run. For example, high-risk technologies may be neglected if firms are uncertain that they will reap the benefits from investing in them. Workers' skills, too, may be subject to market failure. An employer may fear that a worker who receives a training benefit will leave employment before the training pays off for the employer, while the worker may lack the financial means to pay for the training on his or her own. Federal investments in research, technology, and education and training have helped to create and accelerate new industries, such as the semiconductor industry, in the past, when market forces alone would not have done so.

The traditional U.S. approach to innovation has particularly emphasized Federal investments in basic research. This kind of investment is an effective response to an important and continuing market failure. It continues to pay extraordinary dividends, creating opportunities for technological advances in support of U.S. economic growth and national security. The U.S. leads the world in science, and it must continue to strive to maintain this status.

However, the payoffs from Federal investments in basic research have not been fully captured by U.S.-based advanced manufacturing production facilities. In part because of the national strategies of our competitors and in part because of the increasing complexity of industrial technology, private investments in advanced manufacturing capabilities may not occur domestically unless the public sector makes strategic investments to address market failures in stages of the innovation process downstream from basic research. Many of these public investments must be coordinated with private co-investments to create assets such as worker skills, precompetitive technologies, and shared infrastructure, as the President's Council of Advisors on Science and Technology has emphasized.[17]

Therefore, one core principle of an effective national strategy for advanced manufacturing is to take a cohesive approach to research, development, and deployment. The approach begins with more effective commercialization of federally-financed technologies developed as a result of basic research by national labs and universities. More fundamentally, it involves a stronger Federal emphasis on R&D that improves manufacturing processes and supports scale-up, in part by enabling better access to user facilities equipped with advanced manufacturing technologies. In addition, education and workforce training programs that help equip Americans to become highly-skilled manufacturing workers can be invaluable.

Strengthening the Industrial Commons

A cohesive approach to research, development, and deployment is particularly important to encourage investment in high-impact advanced manufacturing processes by small- and medium-sized enterprises (SMEs). SMEs comprise 86% of all manufacturing establishments and employ 41% of the U.S. manufacturing workforce, but they often lag in adopting new technologies.[18] Those SMEs that are highly innovative are often tightly linked to geographically-concentrated communities, or "industrial clusters." These clusters contain other innovative SMEs, larger firms that typically rely on smaller suppliers, academic and training institutions, and other supporting organizations.

Clusters sustain what innovation experts have called the "industrial commons."[19] Like the common pasture in medieval English villages in which livestock owned by many residents grazed together, the industrial commons provides many of today's manufacturers, particularly SMEs, with a chance to refresh their technological base from a set of shared knowledge assets and physical facilities. These common resources help to accelerate innovation and subsequent market penetration. Standards for system interfaces, measurement and test methods, and process control systems, for instance, allow firms within a supply chain or even firms that compete with one another to align their diverse product and process capabilities with opportunities to serve customers in different markets. Similarly, platform technologies in such areas as nanomaterial processing, additive manufacturing, advanced robotics, "smart" manufacturing, and green chemistry are assets that many firms in an industrial cluster can take advantage of, but that no single firm can typically produce on its own.

The industrial commons itself must be refreshed through continual investment to keep the knowledge base and physical infrastructure at the leading edge of technology. It may be difficult for the firms that would benefit from them to make these investments individually, since they cannot capture all of these benefits. Coordination of investments across firms is also subject to market failure. Such investments may also be perceived as too risky for the private sector to make because the gestation periods are long. The risk is heightened in many cases by the prospect that government-linked entities in other countries may make similar investments intended to benefit competing firms. The public sector, particularly Federal agencies, therefore has an important role to play as a co-investor with advanced manufacturers in the industrial commons.

Greater public co-investment in the industrial commons would yield a variety of benefits. It would make basic research investments more productive for the economy. It would strengthen the skills of workers who are in or may enter the advanced manufacturing workforce. And it would reduce private investment risk in product-specific applied R&D, leading to higher rates of technology adoption and more diverse product-specific innovations.[20]

Optimizing Federal Investments

A national strategy for advanced manufacturing should coordinate Federal investments across agencies more effectively. Currently, Federal investments in advanced manufacturing-related research, development, and deployment are largely provided through agency programs focused on accomplishing mission-specific goals. The resulting cross-agency portfolio may not adequately consider how certain investments would benefit multiple agencies and industries as well as contribute to economic competitiveness more generally. A *whole-of-government* innovation policy that complements the work of individual agencies and takes a portfolio view of advanced manufacturing investments would address this problem.

An effective national strategy for advanced manufacturing should also be responsive to private-sector needs. Public-private consultation about research, technology, and workforce needs at the national level will complement the consultative processes that already exist at the state and regional level through-out the country. The President's Advanced Manufacturing Partnership (AMP) represents an initial but important step toward improving private-sector engagement at the national level.

Stakeholders and Objectives

The stakeholders in this national strategy include Federal agencies represented in the Interagency Working Group that authored this report and those represented on the Defense Production Act Committee (DPAC); [21] state, regional, and local public and private entities that support industrial clusters and associated partnerships; manufacturing enterprises of all sizes; a diverse set of institutions of higher education including research universities and community colleges; workers and unions; and the general public.

The Federal strategy for advanced manufacturing seeks to achieve five objectives, which are developed in more detail in the following five sections of the report. These objectives are interconnected; progress on any single one will make progress on the others easier. Coordination of Federal agencies through the NSTC will support an implementation process that pursues all objectives of the strategy in parallel.

Objective 1: *Accelerate investment in advanced manufacturing technology, especially by small and medium-sized manufacturing enterprises, by fostering more effective use of Federal capabilities and facilities, including early procurement by Federal agencies of cutting-edge products.*

Objective 2: *Expand the number of workers who have skills needed by a growing advanced manufacturing sector and make the education and training system more responsive to the demand for skills.*

Objective 3: *Create and support national and regional public-private, government-industry-academic partnerships to accelerate investment in and deployment of advanced manufacturing technologies.*

Objective 4: *Optimize Federal investment in advanced manufacturing by taking a portfolio perspective across agencies and calibrating accordingly.*

Objective 5: *Increase total U.S. public and private investments in advanced manufacturing R&D.*

4. Accelerating Investment by Small and Medium-Sized Enterprises

Objective 1: *Accelerate investment in advanced manufacturing technology, especially by small and medium-sized manufacturing enterprises, by fostering more effective use of Federal capabilities and facilities, including early procurement by Federal agencies of cutting-edge products.*

This component of the strategy aims to improve the success of U.S.-based manufacturers, especially SMEs, in the commercialization and scale-up phases of the technology lifecycle. These phases are critical because they set firms on a path of sustainable job creation and profit generation. The focus on SMEs reflects their importance to the manufacturing sector and the difficulty they experience developing and adopting technological innovations.

We focus on three kinds of Federal actions that support advanced manufacturing investments: (a) increasing coordination of their investments related to advanced manufacturing with private and non-Federal investors, (b) purchasing products made by advanced manufacturers early in the scale-up phase, and (c) investing in targeted areas of critical importance to national security.

Private-Public Co-Investment

Economists have long known that investment opportunities that would benefit groups of firms are less likely to receive private-sector funding than opportunities that benefit individual firms. Investments that would strengthen the industrial commons by deepening an industry's knowledge base or creating industry-wide standards, for instance, are usually less attractive than those that promise to benefit a firm's bottom line directly. Yet, the former are necessary to achieve the latter. These challenges have been exacerbated by recent trends in advanced manufacturing, such as the rising complexity of global markets and the speed with which they change. Rising complexity expands the number of fields that have to be integrated to advance the state of the art, while the faster speed of change raises the risk that ideas and standards will become obsolete before they are fully implemented.

Joint public–private strategic planning and prioritization of opportunities to strengthen the industrial commons should help to overcome these challenges. Coordinating public and private investments related to advanced manufacturing can lead to benefits for both sectors, for instance, by ensuring that all key players participate in standard-setting and accelerating standards adoption.[22] Federal investments in applied research, and in demonstration facilities to which manufacturers have access, can accelerate commercialization of new advanced manufacturing processes and stimulate private investment in plant and equipment.

The President's Advanced Manufacturing Partnership (AMP), announced in June 2011, is an important building block for this element of the strategy. AMP's mission is to identify opportunities for investments in R&D, precompetitive collaboration, and shared facilities and infrastructure that have the potential to transform advanced manufacturing in the United States.

Early Procurement

The Federal government is a major purchaser of products made by advanced manufacturers. Effective use of this purchasing power in support of agency missions can drive economies of scale and scope, especially for SME manufacturers who are then enabled to enter new markets and compete effectively on an international basis. The best-known historical example of effective early procurement is the semiconductor industry. Military buyers absorbed virtually the entire output of semiconductors in 1962. These large-scale purchases pushed prices down 96 percent in the next six years, driving a rapid increase of adoption by the commercial sector, which had grown to be more than 60 percent of the market over that period.[23]

Federal agencies are using early procurement in a variety of mission domains today. In energy, for example, manufacturers of such innovative products as biofuels, advanced batteries, and plug-in electric vehicles, many of them SMEs, are gaining scale economies and production experience as a result of early procurement. But there are substantial opportunities for key agencies such as the General Services Administration and the Department of Defense to expand their use of early procurement in order to accelerate innovation in advanced manufacturing.

Advanced Manufacturing for National Security

The authorities of the interagency Defense Production Act Committee (DPAC) may be drawn upon to strengthen shared infrastructure that supports advanced manufacturing by SMEs. The DPAC is charged with both evaluating current manufacturing capabilities essential to national defense and advising the President on how to resolve manufacturing capability gaps through use of Defense Production Act authorities, such as Title III. These authorities allow Federal investments (usually matched 1:1 with industry) to scale up economically viable means of production (see Appendix B).

Additionally, the Department of Defense (DOD) Manufacturing Technology (ManTech) Program enhances defense industrial base productivity. ManTech's advanced manufacturing investments are applied to drive down weapon system cost and delivery time as well as enhance system performance. This program contributes to improving domestic firms' production effectiveness at every tier of the supply chain. The ManTech program, which has a portfolio of projects in the Army, Navy, Air Force, Defense Logistics Agency (DLA) and Office of the Secretary of Defense (OSD), provides a crucial link between invention and deployment. These investments also enhance manufacturing efficiency and reduce industry's financial and technical risks. ManTech investments currently focus on electronics, metals, composites and advanced manufacturing enterprise (see Appendix C).

Metrics and Key Implementing Agencies for Objective 1

Metrics for short-term progress on Objective 1 could include: (a) documented instances of commercialization and scale-up by firms participating in Federal programs, both manufacturers as a whole and SME manufacturerss (b) use of Federal user facilities by advanced manufacturing firms, and (c) industry matching of Federal DPA Title III investments.

Suggested metrics for long-term progress on Objective 1 include: (a) growth of advanced manufacturing production capacity overall and by SMEs, (b) share of SMEs reporting innovation activity in the manufacturing sector as measured by the Business R&D and Innovation Survey (BRDIS) of the National Science Foundation (NSF),[24] and (c) growth in the U.S.-produced share of the global advanced technology products market.

The primary Federal agencies that should implement actions under Objective 1 include the Departments of Commerce, Defense, Energy, Homeland Security, and Transportation, and the General Services Administration (GSA), National Aeronautics and Space Administration (NASA) and the Small Business Administration (SBA).

5. Strengthening Workforce Skills

Objective 2: *Expand the number of workers who have skills needed by a growing advanced manufacturing sector and make the education and training system more responsive to the demand for skills.*

Unskilled labor was once the mainstay of the manufacturing labor force. As advanced manufacturing supersedes traditional manufacturing, and domestic manufacturers deepen their investment in advanced technologies, the skill requirements for manufacturing jobs are rising. Manufacturing employers perceive a skills gap: 67% of companies surveyed recently by an industry association reported moderate to serious shortages in the availability of qualified workers, even in a period of elevated general unemployment. Certain sectors, such as aerospace/defense and life sciences/medical devices, reported much higher levels of skilled-worker shortages.[25]

Education and training that anticipates and satisfies the skill requirements of advanced manufacturers, while remaining broadly consistent with long-term projections of labor demand, is a key component of this national strategy. Increasing the private sector's confidence in the availability of a skilled advanced manufacturing workforce creates incentives for domestic investment (see Objective 1). These programs should be targeted particularly toward the workforce needs of SMEs. As more advanced manufacturing technology is deployed, on-the-job training becomes more expensive and difficult for companies, especially SMEs, to provide.

Federal actions under this objective should include such efforts as (a) support for the coordination of state and local education and training curricula with advanced manufacturing skill-set requirements, and (b) expanded support for advanced manufacturing career and technical education programs spanning secondary and postsecondary levels, and apprenticeship opportunities through regional partnerships and industrial cluster programs.

The Changing Manufacturing Workforce

The shift from traditional to advanced manufacturing is occurring in the context of a substantial shift in the demographics of the manufacturing workforce. Approximately 2.8 million manufacturing workers (nearly 25%) are now 55 years of age or older.[26] The need to replace these workers as they retire may add to emerging demand for advanced manufacturing workers. In the long term, education and training programs must span from "cradle-to-career" and be responsive to the skill demands of advanced manufacturing employers. Federal programs in cooperation with state and local partners should target (a) separating military personnel and recent veterans, unemployed workers, and employed workers needing to augment their skills in the near-term, (b) prospective workers who will soon enter the work-force, and (c) K–12 students to proactively develop the next-generation of workers.

Better Training for Today's Advanced Manufacturing Workers

The Federal Government is already seeking to adjust current programs that assist state and local public and private efforts to develop and maintain a competitive workforce for advanced manufacturing. Relevant agencies may prioritize advanced manufacturing within workforce development grant programs. They may also identify and disseminate workforce development "best practices" for advanced manufacturing. Many of these practices arise from competitive grants funded by the Department of Labor's Employment and Training Administration (DOL/ETA). Such efforts should be expanded. Among the programs that could increase priority for advanced manufacturing are H-1B Technical Skills Training Grants,[27] the Jobs and Innovation Accelerator Challenge, and Trade Adjustment Assistance Community College and Career Training (TAA/CCCT).

This approach could be complemented by a new emphasis on advanced manufacturing in the promotion of secondary-postsecondary career pathways[28] by the Department of Education's Office of Vocational and Adult Education (ED/OVAE). The President's Fiscal Year (FY) 2013 Budget proposes $8 billion for the Departments of Education and Labor to support state and community college partnerships with businesses to build the skills of American workers in growing industries, such as advanced manufacturing.

Education and Training for Tomorrow's Workers

State and local vocational and apprenticeship training programs supported by the Federal Government strengthen workers' skills. For example, the National Institute of Standards and Technology (NIST) Manufacturing Extension Partnership (MEP) and DOL/ETA support public–private partnerships that establish registered apprenticeship programs in advanced manufacturing. With appropriate input from industry and professional associations, DOL/ETA will seek to ensure that registered apprenticeship programs target needs and gaps in today's advanced manufacturing workforce.

The Advanced Technological Education (ATE) program of the National Science Foundation supports community colleges working in partnership with industry, economic development agencies, workforce investment boards, and secondary and other higher education institutions to respond to industry needs for highly qualified manufacturing technicians. Since the inception of the program in 1994, 265 manufacturing awards have been made totaling $205 million. ATE projects and centers are educating technicians in a range of fields, including nanotechnologies and microtechnologies, rapid prototyping, biomanufacturing, logistics, and alternative fuel automobiles. More details on these and other advanced manufacturing workforce programs can be found in Appendix D.

Another set of partnerships that is being leveraged to improve the training of tomorrow's workers are those organized by the National Association of Manufacturers (NAM) and the Society of Manufacturing Engineers. NAM has helped foster the Manufacturing Skills

Certification System, a broad-based partnership of national organizations seeking to establish a set of credentials that apply to all subsectors of manufacturing.[29] This system draws on the advanced manufacturing competency model developed by the Department of Labor, which identifies the knowledge and skills required to be effective in advanced manufacturing occupations (see Appendix E). Credentials issued within this framework will be nationally portable and industry-recognized. They will help to ensure a smart, safe, and sustainable advanced manufacturing workforce. The Society of Manufacturing Engineers provides extensive training opportunities and associated certifications under this partnership. In June 2011, President Obama drew on the work of NAM, the Society of Manufacturing Engineers, and others when he announced an effort to help 500,000 community college students obtain credentials for advanced manufacturing as part of the Administration's Skills for America's Future initiative.

Educating the Next Generation

A strong science, technology, engineering, and mathematics (STEM) emphasis is needed to prepare students for a variety of post-secondary educational options and a wide range of career opportunities, including careers in advanced manufacturing.[30] Unfortunately, many students who are inclined toward careers as advanced manufacturing technicians who seek to complete a career and technical certificate or community college degree have not received STEM coursework sufficient to succeed in today's advanced manufacturing environment. The President's Educate to Innovate campaign aims to improve participation and performance in STEM education in partnership with leading companies, foundations, and scientific and professional societies.

The perception among some that careers in manufacturing are unattractive and unstable discourages some talented students from seriously exploring them. Federal agencies should consider fostering efforts that aim to transform this perception and consider supporting development of communications materials that accurately depict the opportunities and excitement of 21st century manufacturing.

In recent years, there has been a rapidly growing grassroots movement of "Makers" who are engaged in "do it yourself" projects involving electronics, 3-D printing and robotics. These hands-on projects inspire young people to excel in STEM and can also get them interested in advanced manufacturing. Some agencies, such as DARPA through its Mentor program (see Appendix D), are supporting the Maker movement. More agencies should support Making, in collaboration with the private sector, non-profits, foundations and skilled volunteers.

Another major issue related to preparing students for advanced manufacturing careers is the need to supplement traditional academic education with the development of applied expertise. The Federal Government should help state and local efforts to develop this applied expertise by supporting new manufacturing pre-apprenticeship programs, strengthening existing educational partnerships between community colleges and local industry, and other measures.

Feedback from these programs will help align the curricula of feeder high schools and adult education programs within the service area of each community college. Alignment of these curricula with four-year degree curricula will help provide more attractive career pathways in advanced manufacturing. A critical example is the pathway from a two-year degree to an eventual four-year degree, which enables highly-skilled technical workers to pursue additional education and higher-paying careers.

Metrics and Key Implementing Agencies for Objective 2

The following metrics may be applied to assess progress on Objective 2 in the short-term: (a) industry partnerships with state and local career and technical education providers to specify and implement advanced manufacturing workforce skill sets, and (b) number of individuals earning industry recognized-credentials through career and technical education, vocational training, and apprenticeships programs.

Suggested metrics for assessing long-term progress on Objective 2 include: (a) employment placement rates for graduates from state and local career and technical education, vocational training and apprenticeship programs, and (b) employment levels in highly-skilled manufacturing occupations.

The primary Federal agencies that should implement actions under Objective 2 include DOL/ETA, ED OVAE, and NSF.

6. Creating Partnerships

Objective 3: *Create and support national and regional public-private, government-industry-academic partnerships to accelerate investment in and deployment of advanced manufacturing technologies.*

The acceleration of innovation for advanced manufacturing requires bridging a number of gaps in the present U.S. innovation system. Academic researchers working on problems of importance to advanced manufacturing must communicate more effectively with their counterparts in industry. Federal investments in advanced manufacturing technologies and capabilities must align more fully with similar investments by states and regions and by the private sector. Partnerships among diverse actors, varying by location and objective, are a keystone of our strategy to bridge these gaps.

Actions under this objective will focus more of the Federal advanced manufacturing investment portfolio on partnership activities, especially those that yield benefits for SMEs. These actions include (a) facilitating SME engagement through partnerships, and (b) expanding investments in public–private partnerships in the advanced manufacturing industrial commons.

SME Engagement through Partnerships

SMEs rely on the industrial commons to achieve economies of scale and economies of scope during commercialization and scale up. The production infrastructure that resides in the commons helps SMEs meet diverse market demands through "mass customization" capabilities that use advanced manufacturing technologies to rapidly adjust product attributes while maintaining high quality and low unit cost. Industrial clusters and industry segments that can draw upon a vibrant commons have lower entry barriers for SMEs because these SMEs have easier access to widely-available expertise in advanced manufacturing technologies, innovation, and commercialization.

Federal investments should be better targeted to strengthen the industrial commons, in large part by supporting cross-sectoral partnerships. Partnerships that involve academic institutions, manufacturers, industry associations, and supporting organizations may be structured to produce ideas and capabilities that support commercialization and scale-up activities across a wide swath of firms. A focused strategy for the commercialization of technologies that draw on Federally-funded research, which was recently promulgated by the Administration, will help as well.[31]

The National Design Engineering and Manufacturing Consortium (NDEMC) is a good example of a project that strengthens the industrial commons. Funded by the Economic Development Administration in the Department of Commerce (along with other Federal agency, private, state, and university partners that provide financial or technical assistance), NDEMC enables SMEs to develop and test their products using advanced modeling and simulation tools that have historically only been available to large companies. NDEMC is designed to exploit regional commonalities among supply chain industries and its operational framework is expected to be self-sustaining, scalable, and transferable to other industry sectors and regions across the U.S.

Manufacturing Demonstration Facilities co-funded by DOE with private partners provide physical and virtual tools for rapidly prototyping advanced materials and manufacturing processes within targeted technical areas. These facilities host a collaborative, shared infrastructure that facilitates rapid dissemination of these new technologies. The technical areas already targeted by DOE through this program include carbon fiber and additive manufacturing, with additional areas envisioned in DOE's current plans. MDFs support regional manufacturing ecosystems by providing small- and medium-sized enterprises access to innovative tools and resources that would otherwise be cost-prohibitive and help reduce energy intensity, create lower cost production pathways, and enhance the competitiveness of U.S. advanced manufacturing industries.

Cluster-Based Partnerships

Regional industrial clusters provide a fruitful setting for investments in partnerships. These clusters dramatically improve technology platform development, diffusion of knowledge, and subsequent scale-up. Additional synergies achieved through regional clusters include coordinated strategic planning, complementary asset sourcing and risk pooling within clusters, and co-located supply chains. Clusters provide vital resources for SMEs. Shared production facilities, for example, provide SMEs with both production experience and the knowledge to develop scale-up plans.[32]

The nano-technology cluster that is emerging in Albany, New York, provides an example of the benefits of such partnerships. The State of New York assembled key stakeholders, co-invested with them, and anchored the effort at the State University of New York in Albany. Large manufacturers, such as IBM, along with the semiconductor industry consortium

Sematech, made major commitments, and the cluster has given birth to a number of promising start-ups.[33]

The Federal Government is making cluster-based investments that bring together educational and research organizations, state and regional economic development authorities, and the private sector to conduct proof-of-concept and commercialization activities. For instance, the Jobs and Innovation Accelerator Challenge, which is led by the Economic Development Administration (EDA) in the Department of Commerce in partnership with other agencies, such as the Department of Energy (DOE), NIST, and the Small Business Administration (SBA), will run a competition focused on advanced manufacturing in fiscal 2012.

Federal agencies will also seek to support novel approaches in this context to enhance competitiveness in advanced manufacturing. The E3 Initiative, a framework that connects EPA, NIST, DOE, DOL, and SBA to local and regional groups of SMEs to deploy advanced manufacturing processes to reduce waste and improve operational efficiency, is an example of such a novel approach.[34]

Metrics and Key Implementing Agencies for Objective 3

The following metrics may be used to measure short-term progress on Objective 3: (a) number of Federally-funded public-private partnerships that aim to accelerate innovation in advanced manufacturing, (b) total scale of investment in such partnerships and (c) ratio of private to public funding within these partnerships.

Suggested metrics for long-term progress on Objective 3 include: (a) publications and patents related to advanced manufacturing generated by basic research activities associated with partnerships,[35] (b) number of products commercialized by U.S.-based firms engaged in advanced manufacturing partner-ships and total global sales of such products, and (c) advanced manufacturing location quotients for regions in which advanced manufacturing partnerships are active.[36]

The primary Federal agencies that should implement actions under Objective 3 include DOD, DOE, EDA, NIST, NSF, and SBA.

7. Coordinating Federal Investments

Objective 4: *Optimize the Federal advanced manufacturing investment by taking a portfolio perspective across agencies and adjusting accordingly.*

A number of Federal agencies make research, development, and deployment investments that directly or indirectly benefit advanced manufacturers in the United States. These investments are typically made by agencies independently pursuing their statutory missions. The benefits from these investments can be augmented by analyzing them as a portfolio and adjusting agency investment strategies to reflect this analysis. Such adjustments can be made without compromising agencies' responsiveness to their individual missions. In fact, more efficient and faster development of new technology platforms will enhance the achievement of agency missions.

Actions under this objective include: (a) coordinating Federal agency investments in the industrial commons, and (b) targeting and balancing investments in advanced materials, broad production technology platforms, advanced manufacturing processes, and design and data infrastructure.

The Federal Advanced Manufacturing Investment Portfolio

The Federal Government makes investments in advanced manufacturing R&D and in some kinds of plants and equipment that private industry typically avoids. These high-risk investments help to position promising, nascent technologies (1) for broad adoption and commercialization, or (2) to meet essential DPAC-identified national security needs (see Objective 1). These investments can be grouped into the following four categories:

- Advanced Materials
- Production Technology Platforms
- Advanced Manufacturing Processes
- Data and Design Infrastructure

These four categories encompass the entire spectrum of relevant advanced manufacturing technology investments, from materials that will go into the products of the future to the factory and enterprise processes that enable production. Examples of current Federal investments in these areas are included in Appendix F. Descriptions of the four categories follow. Creating a coordinated Federal portfolio of investments across these four categories will increase the global competitiveness of U.S. manufacturing and help to create a fertile domestic environment for innovation.

Advanced Materials

Materials are the building blocks of every physical product. Traditional materials such as steels, metals, plastics, and ceramics have been improved routinely by materials scientists and chemists. These improvements were vital to many of the significant technological developments of the last century. Newer nanoscale, biological, smart, and composite materials will enable technological breakthroughs in the coming century. Some of these breakthroughs will transform existing industries. Others will spawn entirely new industries.

The Materials Genome Initiative is a key component of the Federal advanced manufacturing investment portfolio. The Departments of Energy and Defense, along with NSF, NIST, and private, academic, and state partners, seek through this initiative to cut the time it takes to develop advanced materials, such as those needed to make vehicles much lighter or dramatically increase the energy density of batteries. The initiative is building the infrastructure and providing the training needed for Americans to discover, develop, manufacture, and deploy advanced materials in a more expeditious and economical way.

Product Technology Platforms

Product technology platforms are the basic means of making products. Manufacturing engineers developed critical improvements in production technologies for components such as display screens, computer chips, batteries, nanoelectronic devices, gears, cases, and bolts. These improvements enabled critical advancements in diverse families of products, such as cell phones, computers, televisions, air bags, automobiles, and satellites. Future product technology platform innovations will enable more flexible and smarter production operations.

Sustainable nanomanufacturing, a signature initiative of the multi-agency National Nanotechnology Initiative, exemplifies Federal efforts to support development of product technology platforms that will provide the basis for future manufacturing industries. These industries will need methods to efficiently assemble products that integrate billions of nanoscale devices with disparate functions. Current manufacturing methods such as those used in the semiconductor industry will not be economical at these scales. The initiative is therefore pursuing radically new technical approaches that promise to lead to flexible, "bottom-up" or "topdown/bottom-up" continuous assembly methods.

Advanced Manufacturing Processes

Manufacturing processes turn materials into components and components into products. Manufacturing process innovations such as semiconductor wafer fabrication, composite structure processing, and bio-production of pharmaceuticals have enabled the introduction of new components and products, such as smartphones, military aircraft, and flu vaccines, in both new and existing markets. Emergent examples of advanced manufacturing processes include additive manufacturing, composite structure manufacturing, and bio-manufacturing.

The Innovative Manufacturing Initiative of the Department of Energy is one element of the Federal portfolio that is seeking to accelerate advanced manufacturing process innovation. It is funding cost-shared R&D of processes that have the potential to significantly reduce energy and carbon intensity over the coming decades. By doing so, the initiative aims to revitalize existing manufacturing industries as well as support the development of emerging industries. In a similar fashion, the Biorefinery Assistance Program of the Department of Agriculture is furthering the development and commercialization of advanced biofuels by providing loan guarantees to first-of-a-kind advanced biofuel biorefineries.

Data and Design Infrastructure

The ability to integrate all three types of innovation described above – advanced materials, product technology platforms, and advanced manufacturing processes – requires effective use of large quantities of manufacturing data and sophisticated design knowledge. Decisions that draw on such data and knowledge to achieve this integration can reduce the time for process improvements to be instituted and for products to reach customers. Data and design infrastructure, such as modeling standards for products and processes, interface standards for functional and physical connections between components, and mechanisms for flexible system integration, make integrated decision-making of this sort far easier and more accessible, especially for SMEs.

The manufacturing initiative of the Defense Advanced Research Project Agency (DARPA) within the DOD includes a number of elements that will strengthen the advanced manufacturing data and design infrastructure. For instance, the initiative seeks to develop design tools that enable correct-by-construction design through model-based verification for complex cyber-physical defense systems. It also seeks to create a bit-stream-configurable "foundry" capable of rapid switch-over across product variants and families leading to mass customization.

Cross-Cutting Agency Investments

Agencies can leverage their funding to invest in projects that they may not otherwise have the resources to support on their own by collaborating with other agencies on joint funding solicitations and co-funding of projects. These cross-cutting manufacturing investments should be used to strengthen the industrial commons in ways that would benefit all the participating agencies (and their stakeholders). For example, agencies could co-fund advanced materials and new design methods that dramatically decrease time to market for innovations. These targeted elements are typically beyond the purview of any one agency or private-sector entity, but are collectively viewed as critical to advancing key national interests. Individual agencies may supplement these investments with additional investments that focus specifically on their agency-unique missions or needs. The recently created Advanced Manufacturing National Program Office provides an administrative mechanism to coordinate and manage such investments.

Metrics and Key Implementing Agencies for Objective 4

Progress toward Objective 4 in the short-term might be measured by: (a) development and implementation of a framework for managing the whole-of-government portfolio, and (b) number and scale of multi-agency advanced manufacturing funding solicitations.

Suggested metrics for long-term progress on Objective 4 include: (a) balance of Federal advanced manufacturing R&D investment across portfolio dimensions, including character of work (i.e. basic research, applied research, demonstration facilities, etc.) and (b) accelerated time-to-market of new advanced manufacturing processes and products.

The primary Federal agencies implementing actions under Objective 4 should include DOD, DOE, NIST, and NSF.

8. Raising National Investment in Advanced Manufacturing R&D

Objective 5: *Increase total U.S. public and private investments in advanced manufacturing R&D.*

Objectives 1 through 4 are essential components of a cohesive national strategy that will increase the payoff from each dollar of Federal investment. They call for leveraging and coordinating Federal investment through partnerships with other levels of government and other public institutions, such as universities, as well as with manufacturers and industry and professional associations. They call for the creation of complementary assets, such as worker skills and industrial know-how, that will fill major gaps in the U.S. national innovation system.

However, actions to meet Objective 1 through 4 will not take full advantage of the present opportunity to sustain and strengthen the Nation's advanced manufacturing sector. To do that, the Nation must raise its level of investment in R&D as well, as the President has called for. The complementary nature of private and public investments suggests that an increase in one sector will be followed by an increase in the other.

Actions under this objective include (a) enhancing and making permanent the Federal Research and Experimentation (R&E) tax credit in order to expand the scope of activities covered and benefit a larger number of manufacturers, and (b) increasing Federal investment for advanced manufacturing R&D.

R&E Tax Credit

Federal tax policy has long provided incentives for private investments in Research and Experimentation (R&E).[37] The policy is well-rooted in economic theory. Private R&D investments create new knowledge that may not pay off for the firm making them, but may turn out to be useful to other firms or to society as a whole, a possibility that may discourage them from making such investments in the first place. The R&E tax incentive helps to overcome this barrier. Further, even applied R&D, which is the focus of industry investment, is risky along both technical and market dimensions compared with other categories of investment. Industry, therefore, systematically under invests in R&D.

However, the structure and size of the current U.S. R&E tax credit is ineffective. Although the U.S. was one of the first nations to enact such a credit, other nations have surpassed the U.S. over the years by offering more attractive provisions. Sixteen industrialized countries that are members of the OECD, for instance, offer tax credits that provide a greater incentive than that of the United States.[38] Such incentives are increasingly important in a world in which there is growing international competition for corporate R&D investment. The President's 2013 Budget proposes enhancing and making permanent the R&E tax credit.

Federal Investment

The Federal Government plays an important role in investing in R&D that helps to foster the development of advanced manufacturing processes and products. The President's Fiscal Year (FY) 2013 Budget provides $2.2 billion for Federal advanced manufacturing R&D—an increase of more than 50% over the 2011 level—at NSF, DOE, NIST, and other agencies.

Important research funded by the Budget includes:

- An additional $86 million above the FY 2012 enacted level for the National Institute of Standards and Technology (NIST) laboratories to expand research in areas such as smart manufacturing, biomanufacturing and nanomanufacturing, as well as $21 million for the Advanced Manufacturing Technology Consortia program, a public-private partnership that will develop road maps for long-term industrial research needs and fund research at universities, government laboratories, and businesses directed at meeting those needs.
- $290 million – more than double the amount in FY 2012 – for the Advanced Manufacturing Office within DOE's Office of Energy Efficiency and Renewable Energy. This program is expanding its R&D activities on innovative manufacturing processes and advanced industrial materials that will enable U.S. companies to cut the costs of manufacturing by using less energy, while improving product quality and accelerating product development. The Budget also continues to support the development of competitive new manufacturing processes for advanced vehicles, biofuels, solar energy and other new clean energy technologies, to help ensure that the technologies invented here are manufactured here.
- An increase of $39 million above the FY 2012 enacted level, for basic research at the National Science Foundation targeted at developing revolutionary new manufacturing technologies in partnership with other Federal agencies and the private sector.

These investments will enhance the competitiveness of the Nation's manufacturing sector by spurring the development of new manufacturing technologies, processes, and materials.

Metrics and Key Implementing Agencies for Objective 5

Potential indicators for measuring progress on Objective 5 over the short term include: (a) enactment of the Administration's tax reforms, including enhancing and making permanent the R&E tax credit, and (b) funding levels for Federal advanced manufacturing R&D.

Metrics for assessing progress on Objective 5 over the long term include: (a) scale of use of the R&E tax credit, and (b) funding levels for Federal advanced manufacturing R&D.

The primary Federal agencies implementing actions should include the Department of the Treasury, DOD, DOE, NIST, and NSF.

APPENDIX A. AMERICA COMPETES REAUTHORIZATION ACT OF 2010 SECTION 102

Coordination of Advanced Manufacturing Research and Development

(a) INTERAGENCY COMMITTEE.-The Director shall establish or designate a Committee on Technology under the National Science and Technology Council. The

Committee shall be responsible for planning and coordinating Federal programs and activities in advanced manufacturing research and development.

(b) RESPONSIBILITIES OF COMMITTEE.- The Committee shall

1. coordinate the advanced manufacturing research and development programs and activities of the Federal agencies;
2. establish goals and priorities for advanced manufacturing research and development that will strengthen United States manufacturing;
3. work with industry organizations, Federal agencies, and Federally Funded Research and Development Centers not represented on the Committee, to identify and reduce regulatory, logistical, and fiscal barriers within the Federal government and State governments that inhibit United States manufacturing;
4. facilitate the transfer of intellectual property and technology based on Federally supported university research into commercialization and manufacturing;
5. identify technological, market, or business challenges that may best be addressed by public-private partnerships, and are likely to attract both participation and primary funding from industry;
6. (encourage the formation of public-private partnerships to respond to those challenges for transition to United States manufacturing; and
7. develop, and update every 5 years, a strategic plan to guide Federal programs and activities in support of advanced manufacturing research and development, which shall-
   A. specify and prioritize near-term and long-term research and development objectives, the anticipated time frame for achieving the objectives, and the metrics for use in assessing progress toward the objectives;
   B. specify the role of each Federal agency in carrying out or sponsoring research and development to meet the objectives of the strategic plan;
   C. describe how the Federal agencies and Federally Funded Research and Development Center supporting advanced manufacturing research and development will foster the transfer of research and development results into new manufacturing technologies and United States based manufacturing of new products and processes for the benefit of society to ensure national, energy, and economic security;
   D. describe how Federal agencies and Federally Funded Research and Development Centers supporting advanced manufacturing research and development will strengthen all levels of manufacturing education and training programs to ensure an adequate, well-trained workforce;
   E. describe how the Federal agencies and Federally Funded Research and Development Centers supporting advanced manufacturing research and development will assist small-and medium sized manufacturers in developing and implementing new products and processes; and
   F. take into consideration the recommendations of a wide range of stakeholders, including representatives from diverse manufacturing companies, academia, and other relevant organizations and institutions,

(c) REPORT.-Not later than 1 year after the date of enactment of this Act, the Director shall transmit the strategic plan developed under subsection (b)(7) to the Senate Committee on Commerce, Science, And Transportation, and the House of Representatives Committee on Science and Technology, and shall transmit subsequent updates to those committees as appropriate.

APPENDIX B. THE DEFENSE PRODUCTION ACT COMMITTEE (DPAC)

The Defense Production Act Committee (DPAC) supports a whole-of-government approach to manufacturing production policy. The DPAC is an interagency body comprised of the majority of the heads of the Federal Government's Departments and Agencies. The Committee's staff conducts assessments of the U.S. industrial base to identify risks within supply chains deemed essential to multiple Department or Agency missions, and provide recommendations to the President for appropriate mitigation. The DPAC also continually reviews DPA authorities and their applications to provide recommendations, as necessary, to Congress and the President on appropriate legal or regulatory modifications.

The DPAC was established by Congress in 2009 to advise the President on the effective use of a law enacted in 1950 (50 U.S.C. App. § 2171 et seq.). The President has directed the Secretaries of Defense and Homeland Security to rotate annually as DPAC Chairperson. DPAC assessments are expected to inform joint-Departmental use of a revolving "DPA Fund," established under Title III of the law to expand the "productive capacity and supply" of important domestic industries. Title III investments are designed to support industrial/technological capabilities that are commercially viable and essential to military production, energy production or construction, military or critical infrastructure assistance to any foreign nation, homeland security, stockpiling, space, and any directly related activity as well as emergency preparedness activities conducted pursuant to title VI of The Robert T. Stafford Disaster Relief and Emergency Assistance Act [42 U.S.C. § 5195 et seq.] and critical infrastructure protection and restoration. Title III assistance is only made available if U.S. industry could not otherwise be expected to provide the needed capability in a timely manner. Moreover, under the law's Statement of Policy in subsection 2(b), "in providing United States Government financial assistance to correct a domestic industrial base shortfall, the President should give consideration to the creation or maintenance of production sources that will remain economically viable after such assistance has ended."

Under Executive Order, the DPA Fund is managed by the DOD, and promotes interagency collaboration with industry on capital expenditures such as retrofits, machine tool acquisitions, or wholesale plant construction. Monies are transferred to the Fund and expended according to requirements established by an Integrated Product Team (IPT), comprised of DOD officials and representatives of the Departments/ agencies contributing to the Fund. According to the DPA Section 2B (1), all Departments and agencies responsible for defense procurement are required to ensure the adequacy of productive capacity and supply. The DPAC supports this key interagency responsibility by providing industrial base assessments and targeted studies of industry areas on behalf of the member Departments and Agencies.

APPENDIX C. THE DEPARTMENT OF DEFENSE (DOD) MANUFACTURING TECHNOLOGY (MANTECH) PROGRAM

The DoD ManTech Program develops advanced manufacturing technologies and processes for the affordable, timely production and sustainment of defense systems. In close partnership with industry, the program impacts all phases of system development, acquisition and sustainment by developing, maturing, and transitioning key advanced manufacturing technologies. Investments are focused on those technologies that have the greatest benefit to the Warfighter and are balanced to support transition of emerging technologies, improvements to existing production enterprises, and strengthening the U.S. Industrial base. ManTech has a long history of delivering critical and "game changing" advanced manufacturing technologies and processes, such as numerically controlled machines, carbon fiber composites, microelectronics fabrication, advanced radars, laser guided munitions, lean production methods, advanced optics, and advanced soldier body armor. Many defense manufacturing technologies have 'spun-off" into commercial markets and resulted in large economic advances for the U.S.

ManTech was established in 1956 and is codified in Title 10, United States Code, to increase national security "...through the development and application of advanced manufacturing technologies and processes that will reduce the acquisition and supportability costs of defense weapon systems and reduce manufacturing and repair cycle times across the lifecycles of such systems." ManTech investments support advanced manufacturing through research and development of processing and fabrication solutions to support emerging technologies as well as enterprise-level initiatives. Current and planned initiatives are described in the 2009 DoD ManTech Program Strategic Plan and include strong support for a highly connected and collaborative manufacturing enterprise, a deep, institutional focus on manufacturing process maturity, and healthy and resilient manufacturing infrastructure and workforce.

The Army, Navy, Air Force, and Defense Logistics Agency (DLA) each manage robust ManTech programs. In addition, the Office of the Secretary of Defense (OSD) has responsibility for managing the Defense-wide Manufacturing Science & Technology (DMS&T) portfolio of R&D investments. The directors and senior managers of these programs coordinate through the congressionally chartered Joint Defense Manufacturing Technology Panel (JDMTP), comprised of one principal representative from each program. The JDMTP identifies and integrates ManTech R&D requirements, conducts joint program planning, and develops joint strategies for advanced manufacturing. ManTech investments are jointly reviewed via a taxonomy of technology focus areas assigned to subpanels. The current subpanel technology focus areas are Electronics, Metals, Composites and Advanced Manufacturing Enterprise.

The position of Deputy Assistant Secretary of Manufacturing and Industrial Base Policy (DASD(MIBP)) was established by the 2011 National Defense Authorization Act (NDAA) and has oversight responsibility for the DoD ManTech Program on behalf of the Secretary of Defense. The DASD(MIBP) is responsible for ensuring that the functions of "manufacturing" and "industrial policy" are effectively integrated and coordinated across the department, and the DoD ManTech Program is an important tool supporting such integration and coordination.

In summary, the DoD ManTech program provides the crucial link between technology invention and robust applications across the U.S. defense industrial base by maturing and validating advanced manufacturing technologies to support low-risk implementation by industry and DoD facilities, such as depots.

APPENDIX D. EXAMPLES OF FEDERAL SUPPORT FOR THE ADVANCED MANUFACTURING WORKFORCE

Defense Advanced Research Projects Agency (DARPA)

DARPA's Manufacturing Experimentation and Outreach (MENTOR) effort engages high school students in collaborative distributed manufacturing and design. MENTOR develops next-generation system designers and advanced manufacturing innovators. MENTOR will (1) develop of user-friendly, open-source tools enabling collaborative distributed design and manufacturing across hundreds of sites and thousands of users through conventional social network media and (2) deploy digitally programmable manufacturing equipment to 1,000 high schools. Clusters of schools will compete in prize-based design and manufacturing challenges to develop of cyber-electromechanical systems of moderate complexity, such as go-carts, mobile robots, or small unmanned aircraft.

Department of Commerce, National Institute of Standards and Technology (NIST)

In June 2011, President Obama announced new commitments by the private sector, colleges, and the Manufacturing Institute of the National Association of Manufacturers (NAM) to train and certify 500,000 community college students through the NAM-Endorsed Manufacturing Skills Certification System. The NIST Manufacturing Extension Partnership (MEP), MEP state centers, and the NAM Manufacturing Institute encourage manufacturers to use the credentialing system in job announcements and in job applicant screening. MEP state centers, the Manufacturing Institute, key workforce partners, and community colleges partner to ensure that their states have the appropriate policies and infrastructures in place for widespread use of the credentialing system to help manufacturers optimize workforce talent.

Department of Education (ED)

The Department of Education *National Career Clusters Framework* supports quality career and technical education programs through learning and comprehensive programs of study (POS). As one of 16 Career Clusters in the Framework, the manufacturing cluster guides development of programs of study in manufacturing that bridge secondary and postsecondary curricula. The manufacturing cluster creates individual student plans for a complete range of manufacturing career options. Career Clusters help students discover their interests and passions and empower them to choose an educational pathway that leads to success in high school, college, and career.

Department of Labor, Employment and Training Administration (ETA)

The Department of Labor's Employment and Training Administration supports the development of a skilled manufacturing workforce through the Registered Apprenticeship program, the Workforce Investment Act programs, and the manufacturing competency model. The Registered Apprenticeship program provides employment and a combination of on-the-job learning with a mentor, technical and theoretical instruction, and progress-driven wage increases. The program meets the needs of high-growth industries by keeping workers' skills updated in response to new technology. Today, there are over 3,000 apprenticeship programs and approximately 17,000 apprentices in manufacturing. The Workforce Investment Act (WIA) programs provide job training and employment services to job seekers and respond to the needs of employers for skilled workers. In the last program year, 19,000 participants who exited the WIA Adult and Dislocated Worker programs received training in occupations in the manufacturing sector. ETA released an updated manufacturing competency model (Appendix E) in spring 2010. Working with industry partners such as the National Association of Manufacturers, the National Council for Advanced Manufacturing, and the Society of Manufacturing Engineers, this employer-validated model outlines the skills necessary to pursue a successful career in the manufacturing industry. The model affords workers in manufacturing fields the ability to advance their training in a way that is consistent with industry demands.

National Science Foundation (NSF)

NSF supports graduate and undergraduate students through its *Graduate Research Fellowship Program* (GFRP) and through relevant competitively awarded research grants. The Advanced Technological Education (ATE) program supports community colleges partnering with industry and others to respond to industry needs for highly qualified manufacturing technicians (see main report text). NSF is creating the next generation of scientific and engineering manufacturing leaders through innovative partnership agreements to provide augmented educational experiences, such as the NSF-wide Grant Opportunities for Academic Liaison with Industry (GOALI), Partnerships for Innovation (PFI), and Accelerating Innovation Research (AIR). These programs provide hands-on educational opportunities for students and/or new entrepreneurs to learn about product development, product qualification, and scale-up of manufacturing.

APPENDIX E. THE ADVANCED MANUFACTURING COMPETENCY MODEL OF THE DEPARTMENT OF LABOR

The competency model framework for Advanced Manufacturing (Figure 3) was developed through a collaborative effort involving the DOL Employment and Training Administration (ETA) and leading industry organizations, including the National Association of Manufacturers, the National Council for Advanced Manufacturing, and the Society of Manufacturing Engineers. To ensure that the model reflects the knowledge and skills needed by today's manufacturing workforce, ETA worked with its industry partners to update the original model. The updated model was completed in April 2010, and contains new information on Sustainable and Green Manufacturing, as well as updated key behaviors in several competency areas.

http://www.careeronestop.org/competencymodel/pyramid.aspx?hg=Y

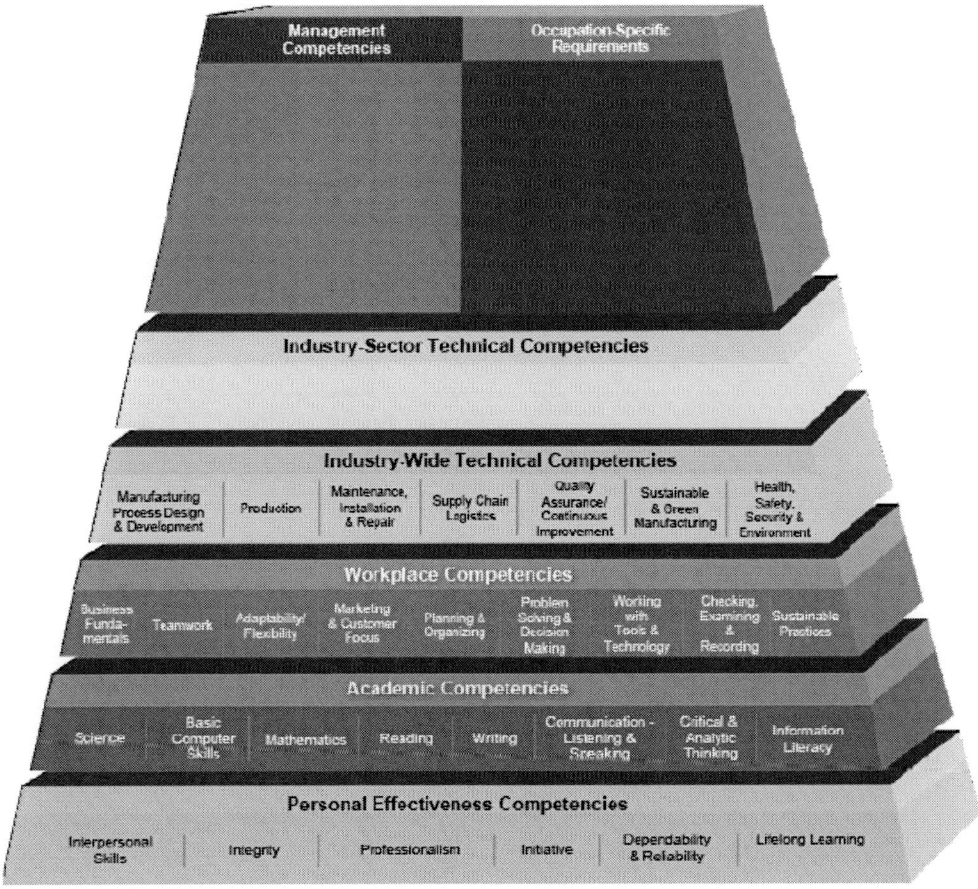

Figure 3. 2010 Advanced Manufacturing Competency Model.

APPENDIX F. EXAMPLES OF FEDERAL INVESTMENTS IN ADVANCED MANUFACTURING R&D

Examples of Federal Investment in Advanced Manufacturing Processes

DOD DARPA	Antibody Technology Program
DOD DARPA	Open Manufacturing
DOD DARPA	Disruptive Manufacturing Technologies
DOD DARPA	HEALICS (Self-HEALing Mixed-Signal Integrated Circuits)
DOD DARPA	Tip-based Nanofabrication
DOD ManTech	Rapid Manufacturing of Aerospace Structures
DOD ManTech	Prosthetics and Orthotics Manufacturing Initiative
DOD ManTech	Joint Strike Fighter Producibility and Affordability Initiatives
DOE EERE	Innovative Manufacturing Initiative – Innovative Manufacturing Processes
EPA	Nanotechnology Research Program
EPA	Green Chemistry Awards
EPA	Energy Star
NASA	Next Generation Robotics
NASA	EBF3 Electron-Beam Freeform Process Fabrication Demonstration
NASA	Strong Cities, Strong Communities
NIST	Next-Generation Robotics and Automation
NIST	Integrated Smart Manufacturing Processes, Equipment, and Systems
NSF	Core Research Programs in Multiple Directorates
NSF	Engineering Research Center Program
NSF	Industry/University Cooperative Research Center Program
NSF	Grant Opportunities for Academic Liaison with Industry (GOALI) Program
USDA	Biomass Research and Development Initiative
USDA	Specialty Crop Research Initiative
USDA	McIntire-Stennis Formula Grants
USDA	Hatch Act Formula Funds, Evan-Allen Formula Funds, Smith-Lever Formula Funds, Extension Programs for 1890 Institutions Formula Funds
USDA	Bioenergy
USDA	Bioenergy for Advanced Biofuels
USDA	Biorefinery Assistance Program

Examples of Federal Investment in Advanced Materials

DOD DARPA	Hybrid Multi-Material Rotor
DOD DARPA	Lightweight Ceramic Armor
DOD DARPA	Structural Logic
DOD ManTech & DPA Title III	Producibility of Large, Affordable Substrates for Advanced Imaging
DPA Title III	Carbon Nanotube Development & Production Enhancement
DOD ManTech	Advanced Body Armor Material Production Technologies
DOD ManTech	Affordable, Advanced Titanium Powder Processing
DOE ARPA-E	BEEST (Batteries for Electrical Energy Storage for Transportation)
DOE EERE	Innovative Manufacturing Initiative – Innovative Materials
NIST	Next-Generation Materials Measurements, Modeling, and Simulation
NSF	Advanced Materials Genome Initiative
NSF	Materials by Design
NSF	Materials Research Science and Engineering Centers

Examples of Federal Investment in Product Technology Platforms

DOD DARPA	Manufacturable Gradient Index Optics
DOD DARPA	IRIS (Integrity and Reliability of Integrated Circuits)
DOD DARPA	GRATE (Gratings of Regular Arrays and Trim Exposures)
DOD DARPA	Maskless Nanowriter
DOD ManTech	Advanced Electronics Packaging and Fabrication
DOD ManTech	Lithium-ion Battery Producibility Advancements
DOE ARPA-E	Direct Wafer Technology
DOE EERE	Innovative Manufacturing Initiative – Innovative Manufacturing Processes
NIST	Biomanufacturing
NIST	Nanomanufacturing
NIST	Advanced Semiconductor Electronics Manufacturing
NIST	Flexible Electronics Manufacturing
NSF	Nanomanufacturing
NSF	Bio-Economy
NSF	Cyber Physical Systems

Examples of Federal Investment in Data and Design Infrastructure

DOD DARPA	ADAPT
DOD DARPA	META
DOD DARPA	iFAB (instant Foundry, Adaptive through Bits)
DOD DARPA	FANG (Fast Adaptable Next generation Ground vehicle)
DOD DARPA	Diverse and Accessible Heterogeneous Integration
DOD ManTech	Connecting American Manufacturing Initiative
DOD ManTech	Advanced Supply Chain Modeling
DPA Title III	Heavy Forge Production Capacity Improvement
DHS	SECURE (System Efficacy through Commercialization Utilization Relevance and Evaluation)
DHS	FutureTECH
DHS	SAFETY Act (Support Anti-terrorism by Fostering Effective Technologies Act)
EPA	Lean Manufacturing
EPA	Green Engineering
EPA	Design for Environment
NIST	Sustainable Manufacturing
NSF	Engineering Research Centers
NSF	Industry/University Cooperative Research Centers
NSF	Innovation Corps
USDA	Quality and Utilization of Agricultural Products
USDA	Forest Products Utilization

End Notes

[1] President's Council of Advisors on Science and Technology, *Report to the President on Ensuring American Leadership in Advanced Manufacturing*, June 2011, p. ii.

[2] This report does not address broader issues such as Federal tax (with the exception of the research and experimentation tax credit), infrastructure, investment, intellectual property, trade, and export-promotion policies that are associated with competitiveness and innovation but that are explicitly assigned in Section 604 of COMPETES to a separate report from the Department of Commerce, which was published in January 2012. See http://www.commerce.gov/americacompetes.

[3] The Census Bureau defines Advanced Technology Products using about 500 of some 22,000 commodity classification codes used in reporting U.S. merchandise trade. Each of the 500 codes meets the following three criteria – (1) the code contains products whose technology is from a recognized high technology field, (2) these products represent leading edge technology in that field, and (3) such products constitute a significant part of all items covered in the selected classification code.

[4] PCAST, *Advanced Manufacturing.*

[5] Bureau of Economic Analysis, 2010 U.S. Economic Accounts by Industry, see http://www.bea. gov/industry/ index.htm

[6] Bureau of Labor Statistics, 2011 *Employer Costs for Employee Compensation,* Table 6.

[7] Bureau of Economic Analysis, *Industry-by-Industry Total Requirements Table*, see http://www. bea.gov/industry/ iotables/prod/

[8] Bureau of Economic Analysis and Census, *U.S. International Trade in Goods and Services.*

[9] National Science Board, *Science and Engineering Indicators* 2012, Appendix Table 4-14 and Table 3-32.

[10] Peter Bisson, Elizabeth Stephenson, and S. Patrick Viguerie, "The Productivity Imperative," *McKinsey Quarterly*, June 2010.

[11] National Science Foundation, *Science and Engineering Indicators* 2012, p. 4-42, see http:// www.nsf.gov/ statistics/ seind12/.

[12] R. Atkinson and S. Andes, *The Atlantic Century II: Benchmarking E.U. and U.S. Innovation and Competitiveness*. Washington, DC: Information Technology and Innovation Foundation, 2011.

[13] The U.S. "major-currency" dollar index is the value (weighted geometric mean) of the dollar relative to a basket of foreign currencies (Euro, English pound, Canadian dollar, Swedish krona, Swiss franc, and Japanese yen). During this same period, the dollar declined 22 percent against an index of all currencies. See http://www.federalreserve.gov/releases/h10/ summary/default.htm

[14] R&D intensity is measured by R&D as a share of industry sales; R&D-intensive industries have R&D/sales ratios of greater than 3%.

[15] Gary P. Pisano and Willy C. Shih (2009), Restoring American competitiveness, *Harvard Business Review,* July.

[16] Secretary of Commerce John Bryson, "Remarks at U.S. Chamber of Commerce," December 15, 2011.

[17] PCAST, *Advanced Manufacturing.*

[18] SMEs are defined as firms with fewer than 500 employees, consistent with the Small Business Administration's definition. Data are drawn from the Business Dynamics Statistics database, U.S. Census Bureau.

[19] Pisano and Shih, *op. cit.*

[20] Tassey, Gregory (2010), "Rationales and Mechanisms for Revitalizing U.S. Manufacturing R&D Strategies", Journal of Technology Transfer 35 (June): 283-333.

[21] The DPAC is described in more detail below and in Appendix C.

[22] OSTP/USTR/OMB Memorandum on "Principles for Federal Engagement in Standards Activities to Address National Priorities," January 17, 2012, http://www.whitehouse.gov/ sites/default/files/omb/memoranda/ 2012/ m-12-8_1.pdf

[23] Morris, Peter Robin, *A History of The World Semiconductor Industry* (London: Peter Peregrinus, 1990), p. 75.

[24] National Science Foundation (2011) *NSF Releases New Statistics on Business Innovation,* NSF11-300, http://www.nsf.gov/statistics/infbrief/nsf11300/nsf11300.pdf

[25] Deloitte Consulting LLP, Manufacturing Institute (2011), *Boiling Point? The skills gap in U.S. manufacturing.*

[26] Bureau of Labor Statistics, *2010 Current Population Survey.*

[27] H-1B Technical Skills Training Grants provide education, training and job placement assistance related to high-growth fields in which employers are currently using the H-1B nonimmigrant visa program to hire foreign workers, including advanced manufacturing, These grants are funded through fees paid by employers under the H-1B program.

[28] These are referred to as "programs of study" in the Perkins CTE legislation.

[29] The Manufacturing Institute (2010) NAM-Endorsed Manufacturing Skills Certification System, see http://www.themanufacturinginstitute.org/Education-Workforce/Skills-Certification-System/Skills-Certification-System.aspx

[30] The National Science and Technology Council's Committee on Science, is carrying out a comprehensive effort to strengthen the Nation's STEM education system. The most recent public report is NSTC, *The Federal Science, Technology, Engineering, and Mathematics (STEM) Education Portfolio (2011).*

[31] Presidential Memorandum (October 2011) *Accelerating Technology Transfer and Commercialization of Federal Research in Support of High-Growth Businesses.*

[32] Gregory Tassey (2011) *Beyond the Business Cycle: The Need for a Technology-Based Growth Strategy* (http://www.nist.gov/director/planning/upload/beyond-business-cycle.pdf). For a broader statement of cluster theory and empirical evidence, see Mercedes Delgado, Michael E. Porter, and Scott Stern, "Clusters, Convergence, and Economic Peformance," Institute for Strategy and Competitiveness, Harvard Business School, March 2011 (http://www.isc. hbs.edu/pdf/DPS_Clusters_Performance_2011-0311.pdf).

[33] Board on Science, Technology, and Economic Policy, National Research Council, Growing Innovation Clusters for American Prosperity: *Summary of a Symposium* (National Academies Press, 2011).

[34] E3: Economy, Energy, Environment: Supporting Manufacturing Leadership through Sustainability, see www.e3.gov.

[35] Due to the nature of basic research, metrics for long term progress on this objective will draw on best assessment practices from existing industry-university research center programs (e.g. NSF Engineering Research Centers, NSF Industry / University Collaborative Research Centers) as well as the use of emerging Science of Science Policy tools (e.g. Star Metrics) as appropriate.

[36] Location Quotients (LQs) are ratios that allow an area's distribution of employment by industry to be compared to a reference or base area's distribution. A high location quotient means that an industry is particularly concentrated in an area. See BLS help page: http:// www.bls.gov/help/def/lq.htm.

[37] The statutory definition of "research and experimentation" differs from the colloquial "research and development" in that it excludes spending that supports development of a specific product. See Science and Engineering Indicators 2012, p. 4-36.

[38] *OECD, Science, Technology and Industry: Outlook 2008*, http://www.sourceoecd.org/ 9789264049918.

In: Advanced Manufacturing in America
Editors: Brandon Cullerne and Michael Laston

ISBN: 978-1-62257-932-7

Chapter 2

CAPTURING DOMESTIC COMPETITIVE ADVANTAGE IN ADVANCED MANUFACTURING*

President's Council of Advisors on Science and Technology

EXECUTIVE REPORT OF THE AMP STEERING COMMITTEE

Advanced manufacturing is not limited to emerging technologies; rather, it is composed of efficient, productive, highly integrated, tightly controlled processes across a spectrum of globally competitive U.S. manufacturers and suppliers. For advanced manufacturing to accelerate and thrive in the United States, it will require the active participation of communities, educators, workers, and businesses, as well as Federal, State, and local governments.

The Advanced Manufacturing Partnership (AMP) Steering Committee proposes that the Nation establish a national advanced manufacturing strategy. This strategy will serve as a national framework that, when implemented by states and local communities, will bring about a sustainable resurgence in advanced manufacturing in the United States.

The AMP Steering Committee developed a set of 16 recommendations around three pillars:

- Enabling innovation
- Securing the talent pipeline
- Improving the business climate

These recommendations are aimed at reinventing manufacturing in a way that ensures U.S. competitiveness, feeds into the Nation's innovation economy, and invigorates the domestic manufacturing base. The objective is to position the Nation to lead the world in new disruptive advanced manufacturing technologies that are changing the face of manufacturing.

---

* This is an edited, reformatted and augmented version of the AMP Steering Committee Report, dated August 8, 2011.

The AMP Steering Committee believes that a number of important steps taken now will be critical to strengthen the Nation's innovation system for advanced manufacturing. While some of the largest U.S. firms have the depth and resources to be ready for this challenge, a significant number of small and medium-sized U.S. firms operate largely outside the present innovation system. The United States will only lead in advanced manufacturing if all companies are able to participate in the transformations made possible through innovations in manufacturing.

The AMP Steering Committee proposes 16 recommendations that will set the stage for advanced manufacturing to thrive in the United States

Enabling Innovation

1. **Establish a National Advanced Manufacturing Strategy:** The AMP Steering Committee proposes establishing and maintaining a national advanced manufacturing strategy by putting in place a systematic process to identify and prioritize critical crosscutting technologies.
2. **Increase R&D Funding in Top Cross-Cutting Technologies:** In addition to identifying a "starter list" of crosscutting technologies that are vital to advanced manufacturing, the AMP Steering Committee proposes a process for evaluating technologies for research and development (R&D) funding.
3. **Establish a National Network of Manufacturing Innovation Institutes (MIIs):** The AMP Steering Committee proposes the formation of MIIs as publicprivate partnerships to foster regional ecosystems in advanced manufacturing technologies. MIIs are one vehicle to integrate many of the Committee's recommendations.
4. **Empower Enhanced Industry/University Collaboration in Advanced Manufacturing Research:** The AMP Steering Committee recommends a change in the treatment of taxfree bond-funded facilities at universities that will enable greater and stronger interactions between universities and industry.
5. **Foster a More Robust Environment for Commercialization of Advanced Manufacturing Technologies:** The AMP Steering Committee recommends that action is taken to connect manufacturers to university innovation ecosystems and create a continuum of capital access from start up to scale up.
6. **Establish a National Advanced Manufacturing Portal:** The AMP Steering Committee recommends that a searchable database of manufacturing resources is created as a key mechanism to support access by small and mediumsized enterprises to enabling infrastructure.

Securing the Talent Pipeline

7. **Correct Public Misconceptions About Manufacturing:** Building excitement and interest in careers in manufacturing is a critical national need, and an advertising campaign is recommended by the AMP Steering Committee as one important step in this direction.

8. **Tap the Talent Pool of Returning Veterans:** Returning veterans possess many of the key skills needed to fill the skills gap in the manufacturing talent pipeline. The AMP Steering Committee makes specific recommendations on how to connect these veterans with manufacturing employment opportunities.
9. **Invest in Community College Level Education:** The community college level of education is the "sweet spot" for reducing the skills gap in manufacturing. Increased investment in this sector is recommended, following the best practices of leading innovators.
10. **Develop Partnerships to Provide Skills Certifications and Accreditation:** Portability and modularity of the credentialing process in advanced manufacturing is critical to allow coordinated action of organizations that feed the talent pipeline. The AMP Steering Committee supports the establishment of stackable credentials.
11. **Enhance Advanced Manufacturing University Programs:** The AMP Steering Committee recommends that universities bring new focus to advanced manufacturing through the development of educational modules and courses.
12. **Launch National Manufacturing Fellowships & Internships:** The AMP Steering Committee supports the creation of national fellowships and internships in advanced manufacturing in order to bring needed resources but more importantly national recognition to manufacturing career opportunities.

Improving the Business Climate

13. **Enact Tax Reform:** The AMP Steering Committee recommends a set of specific tax reforms that can "level the playing field" for domestic manufacturers.
14. **Streamline Regulatory Policy:** The AMP Steering Committee recommends a framework for smarter regulations relating to advanced manufacturing.
15. **Improve Trade Policy:** Trade policies can have an adverse impact on advanced manufacturing firms in the United States. The AMP Steering Committee recommends specific actions that can be taken to improve trade policy.
16. **Update Energy Policy:** The manufacturing sector is a large consumer of energy, and consequently, domestic energy policies can have a profound impact on global competitiveness. The AMP Steering Committee makes specific policy recommendations regarding energy issues of importance in manufacturing.

With sustained focus, alignment of interests, and coordinated action to implement the above recommendations, the United States can and will lead the world in advanced manufacturing. Already today, there are examples of new manufacturing technologies emerging from research laboratories that will have a transformative effect on the way manufacturing is done in America. Together, government, industry, and academia must commit to reinvent the national manufacturing base to ensure our Nation's future.

I. ADVANCED MANUFACTURING PARTNERSHIP

"Advanced manufacturing is a family of activities that (a) depend on the use and coordination of information, automation, computation, software, sensing, and networking, and/or (b) make use of cutting edge materials and emerging capabilities enabled by the physical and biological sciences, for example nanotechnology, chemistry, and biology. It involves both new ways to manufacture existing products, and the manufacture of new products emerging from new advanced technologies."

—President's Council of Advisors on Science and Technology Report to the President on Ensuring American Leadership in Advanced Manufacturing, p. ii

Genesis

The United States has long thrived as a result of its ability to manufacture goods and sell them to global markets. Manufacturing has supported our economic growth, contributing to the Nation's exports, and employing millions of Americans. Manufacturing has driven knowledge production and innovation in the United States by supporting twothirds of private sector research and development (R&D) and by employing scientists, engineers, and technicians to invent and produce new products.[1]

Advanced manufacturing encompasses all aspects of manufacturing, including the ability to quickly respond to customer needs through innovations in production processes and innovations in the supply chain. As manufacturing advances, it is increasingly becoming knowledgeintensive, relying on information technologies, modeling, and simulation. Manufacturers are also increasingly focusing on environmentallysustainable practices that lead to improved performance and reduced waste.

The benefits to focusing on advanced manufacturing are many. As Figure 1 shows, manufacturing creates more value across the economy per dollar spent than any other sector. Manufacturing produces new goods that fundamentally change or create new services and sectors.

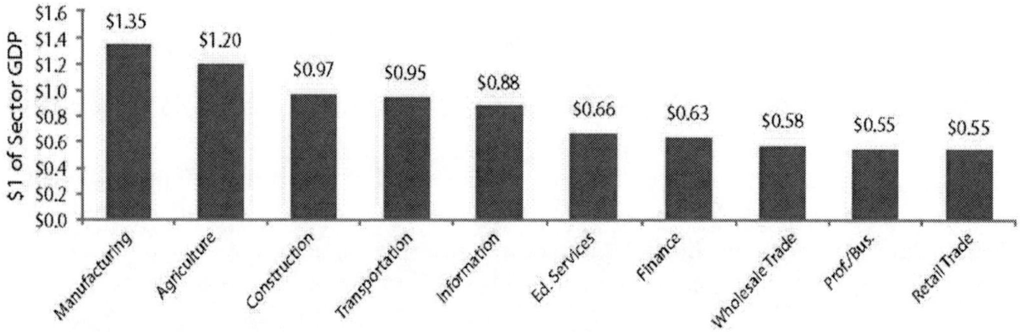

Source: AMP Steering Committee based on data from Bureau of Economic Analysis, Input-Output Tables available at http://www.bea.gov/iTable/index_industry.

Figure 1. Economic Activity Generated by $1 of Sector Output, 2010.

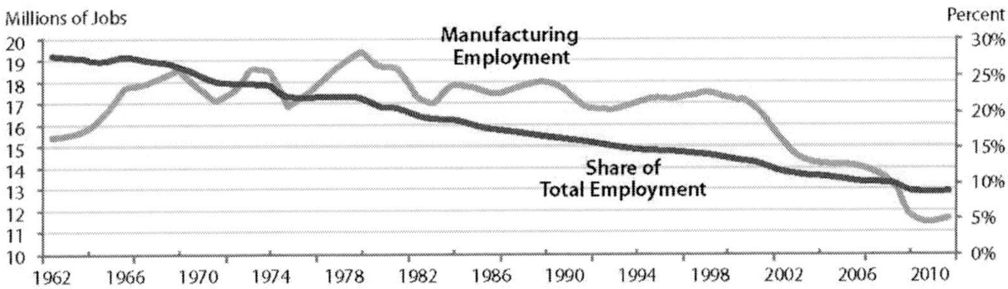

Source: AMP Steering Committee based on data from Bureau of Labor Statistics, Current Employment Statistics, 1962–2010 provided in Table B1 at http://www. bls.gov/ces/tables.htm#ee.

Figure 2. Employment Trends, 1962–2010.

However, the nation's historic leadership in manufacturing is at an inflection point. Although the United States has been the leading producer of manufactured goods for more than 100 years, manufacturing has been declining as a share of U.S. gross domestic product (GDP) and employment (Figure 2).

The loss of U.S. leadership in manufacturing is not limited to low-wage jobs in low-tech industries, nor is it limited to our status relative to low-wage nations. The hard truth is that the United States is lagging behind in innovation in the manufacturing sector relative to high-wage nations such as Germany and Japan, and the United States has relinquished leadership in some medium and high-tech industries that employ a large proportion of highly-skilled workers. In addition, the United States has been losing significant elements of the research and development (R&D) activity linked to manufacturing to other nations, as well as its ability to compete in the manufacturing of many products that were invented and innovated here—from laptop computers to flat panel displays to lithium ion batteries. Recognizing this, in June 2011, the President's Council of Advisors on Science and Technology (PCAST) and the President's Innovation and Technology Advisory Committee (PITAC) issued a report to the President on *Ensuring American Leadership in Advanced Manufacturing*.[2] The report provided a strategy and specific recommendations for revitalizing the Nation's leadership in advanced manufacturing.

To ensure that the United States attracts manufacturing activity and remains a leader in knowledge production, the report recommended the following two strategies:

1. "Create a fertile environment for innovation so that the United States provides the overall best environment for business, through tax and business policy, robust support for basic research, and training and education of a high-skilled workforce; and
2. Invest to overcome market failures, to ensure that new technologies and design methodologies are developed here, and that technology-based enterprises have the infra structure to flourish here."[3]

Based on the PCAST report, on June 24, 2011, President Obama launched the Advanced Manufacturing Partnership (AMP), a national effort bringing together industry, universities, the Federal Government, and other stakeholders to identify emerging technologies with the potential to create highquality domestic manufacturing jobs and enhance U.S. global competitiveness.

Operating within the framework of PCAST, the AMP Steering Committee had three targeted outcomes, which fit intimately together and will have an additive effect when implemented:

1. Develop a permanent model for evaluating, prioritizing, and recommending Federal investments in advanced manufacturing technologies,
2. Recommend a set of partnership projects, focused on advancing highimpact technologies and creating models for collaboration that encompass technology development, innovation infrastructure, and workforce development,
3. Provide recommendations to the Administration on the actions required to support investment in advancing manufacturing in the United States.

Process

After its launch, the AMP Steering Committee initiated five workstreams with the objectives listed in Table 1.

The AMP Steering Committee engaged in extensive consultations with stakeholders across the country to identify opportunities for investments in advanced manufacturing that have the potential to transform U.S. industry. Most notably, four regional meetings were conducted—in Atlanta, Georgia; Cambridge, Massachusetts; Berkeley, California; and Ann Arbor, Michigan—providing a forum for 1,200 attendees, representing industry, academia, and government, to openly share their observations, views, and recommendations. (See Annex 6 for summaries of these meetings.) In addition, extensive surveys were conducted through various manufacturing and academic trade and professional associations.

Table 1. Workstream Objectives

Workstream	Objectives
Technology Development	• Determine a permanent mechanism to be used for identifying and developing key manufacturing technologies • Determine a set of top technology areas that would ensure U.S. manufacturing competitiveness
Shared Infrastructure & Facilities	• Assess opportunities to de-risk, scale-up, and lower the cost of accelerating technology from research to production through unique capabilities and facilities that serve all U.S. based manufacturers, in particular, small and medium-sized enterprises
Education & Workforce Development	• Identify tangible actions to support a robust supply of talented individuals to provide human capital to companies interested in investing in advanced manufacturing activities in the United States
Policy	• Make recommendations to the Administration on economic and innovation policies that can directly impact the overall climate and the ability to improve research collaboration and the pathway to commercialization in support of U.S.-based manufacturing and jobs
Outreach	• Conduct stakeholder outreach and reviews • Conduct and consolidate findings of regional meetings

Source: AMP Steering Committee

These consultations contributed significantly to the recommendations of the AMP Steering Committee to PCAST. The Steering Committee firmly believes that these recommendations will provide the foundation for future breakthroughs by building a national roadmap for advanced manufacturing technologies, speeding ideas from the drawing board to the manufacturing floor, scaling-up first-of-a-kind technologies, training our future workforce, and developing the infrastructure and shared facilities to allow small, mid-sized, and large manufacturers to innovate and compete.

Each work-stream prepared a report, providing the basis for this AMP Steering Committee report and its recommendations. These reports can be found in the following annexes:

- *Annex 1:* Technology Development Workstream
- *Annex 2:* Shared Facilities and Infrastructure Workstream
- *Annex 3:* Education and Workforce Workstream
- *Annex 4:* Policy Workstream
- *Annex 5:* Outreach Workstream

II. ADVANCED MANUFACTURING MATTERS

Role of Advanced Manufacturing in the Global Economy

During the 20th century, U.S. manufacturing increased production at a relatively steady rate. Overall investment in capacity steadily expanded. Manufacturing employment held at roughly 17 to 18 million from the late 1960s through the late 1990s. Over the last decade, however, this equation changed. Production has been nearly flat for over a decade. The United States lost one-third of its manufacturing workforce, and investment in new production capacity stalled.[4] Productivity gains alone cannot be blamed for the loss in manufacturing employment over this period; it is rather the overall loss in the competitiveness of the sector. There are many contributing factors. These losses have led to large trade deficits in manufactured goods, including in advanced technology products, and a loss in global share of manufacturing production. There are growing concerns that the loss in capacity over the last decade has also impacted our domestic innovation and manufacturing capabilities, impeding new investment in domestic manufacturing.

Advanced manufacturing serves a critical role in today's economy. Manufacturing contributes disproportionately to U.S. innovation. Proximity to the manufacturing process creates innovation spillovers across firms and industries, leading to the ideas and capabilities that support the next generation of products and processes. In this way, a vibrant manufacturing sector is inextricably linked to our capacity as a nation to innovate.

Despite recent declines in manufacturing employment, manufacturing industries still employ nearly 12 million workers. These industries are responsible for a significant portion of domestic R&D investment—a key driver of innovation. Small and medium-sized enterprises in the manufacturing sector are a critical component of the U.S. economy, representing 84 percent of manufacturing establishments in 2009[5] and employing 51 percent of the U.S. manufacturing workforce in 2010.[6]

The impact of a healthy manufacturing sector has a ripple effect on the economy. On average, each manufacturing job supports 2.5 jobs in other sectors, and, at the upper end, each high-tech manufacturing job supports sixteen others.[7] Each dollar in final sales of manufactured goods supports $1.35 in output from other sectors of the economy.[8] Compared to all other sectors, manufacturing has the largest multiplier.[9] Manufacturing not only spurs tremendous economic activity, it encourages innovation and research wherever it occurs. In 2009, manufacturing domestic business R&D spending in the *United States reached $195 billion, accounting for 70 percent of all domestic business R&D performed in the United States.*[10]

The importance of manufacturing to employment is not measured by simply counting the numbers of production workers. The production stage affects employment throughout long product value chains, from the innovation and input stages for product design and production including resources, components, suppliers, to the output stages including distribution, sales and the maintenance and repair life cycle service for the product. Total employment for manufacturing, and therefore its economic impact, is much bigger than simply those engaged at the production moment itself.

Manufacturing also has a significant effect on the global trade balance. Over the prior decade, manufactured goods represented 65 percent of U.S. trade.[11] A decline in the U.S. manufacturing base over the last two decades has led, in part, to chronic trade deficits. The United States has, in fact, run a trade deficit in advanced technology products every year since 2002.[12] There is simply no way to reduce these chronic trade deficits without a vibrant manufacturing sector; it is not possible for this deficit to be balanced through the service sector alone.

To reinvigorate the U.S. economy and pursue long-term economic prosperity, America must reject the notion that the Nation should let go of its manufacturing sector in favor of services. No other sector creates more high paying jobs that sustain a vast swath of American households. Instead, the United States needs to recognize that manufacturing and services are interdependent and the success of one sector affects the other. Those in the industry know it is not effective to separate the manufacturing and service sectors; manufacturing and innovation go hand-in-hand. Economic growth will not be sustainable if the two are decoupled. If the Nation attempts to rely on innovation alone, innovation—and the value it creates—will follow manufacturing overseas.

Importance of Manufacturing to National Security

Maintaining technological superiority in advanced manufacturing is a national security issue and is critically important for sustaining U.S. global competitiveness. A strong manufacturing sector not only ensures a ready supply of defense and commercial goods and services, but also ensures the integrity of these goods, especially electronics and other mission critical items. However, U.S. national security is not limited to the products and technologies that are required for national defense; it also entails the products and technologies required for our nation's energy security, food security, heath security, cybersecurity and economic security.

Moving forward, the United States must maintain access to low-cost, secure sources of energy. The Nation has already made great strides in photovoltaics, advanced energy storage devices, and alternate feedstocks, but it needs to accelerate the development of advanced manufacturing technologies to deliver cost competitive economics. The United States, and the world, is witnessing the global importance of food security; advanced manufacturing will enable the world to feed its growing population in a sustainable, efficient manner through high-tech seeds and plant genomics. An aging global population is increasingly relying on cutting edge pharmaceuticals and medical technology, a sector in which advanced manufacturing plays a pivotal role. Finally, every sector of the economy is increasingly dependent on information technology systems; hence, not only is information technology vitally important to national security, but to leadership in the global economy. Breakthrough information technologies require advances in manufacturing to deliver the next generation of systems and tools, and advanced manufacturing is depending on these future systems for next generation processes.[13]

Interplay between Innovation and Advanced Manufacturing

Other countries have witnessed the unparalleled economic prosperity created by a manufacturing economy, and appreciate the inherent value of manufacturing. They are actively competing for manufacturing technologies and manufacturing production. The major economic competitors of the United States recognize the benefits from a vibrant manufacturing sector, and they have developed approaches for attracting manufacturing investment. In doing so, other countries are capturing the R&D that follows the manufacturing. As economist Gregory Tassey of the National Institute of Standards and Technology writes: "The issue of co-location of R&D and manufacturing is especially important because it means the value-added from both R&D and manufacturing will accrue to the innovating economy, at least when the technology is in its formative stages."[14] Many argue that R&D and manufacturing can be separated with the United States focusing on R&D and design. However, studies have shown that off-shoring of manufacturing leads to later loss of R&D competencies. "Losing this [manufacturing] exposure makes it harder to come up with innovative ideas."[15] Related to this argument, building manufacturing plants in the United States has additional benefits of providing quicker access to supplies of intermediate goods and services; access to a larger pool of workers; proximity to consumers; and increased flow of knowledge spillovers across firms through the supply chain and worker mobility.[16]

The problems the world faces are complex. They cannot be solved by services alone. They require the innovation, creativity, and ingenuity of manufacturing companies, together with that of academia and national research laboratories. As the world's population rises and new economies emerge, society requires novel solutions to meet its pressing needs for energy, water, food, health, security, and public infrastructure. Solutions to these challenges are complex and require novel approaches. No longer can the problems be solved by singular disciplines. They require interdisciplinary approaches and collaboration between the private and public sectors. They require partnering among the world's best universities, entrepreneurs, national labs, and small, medium, and large enterprises to address the world's most pressing challenges, uncover scientific fundamentals, discover new molecules and materials, and scale new processes and operations.

U.S. Global Competitiveness

By failing to update and realign its policies, the United States is slowly ceding its position as the long-standing leader in advanced manufacturing. Nations around the world are offering a more positive climate for new industrial plants and to encourage business investment locally. Public policies should encourage investment. World-class educational systems and workforce training practices serve as magnets for manufacturers. In recent years, the R&D system support by the U.S. Government has had very limited focus on technology advances needed for advanced manufacturing. This benign neglect has taken its toll and is in sharp contrast to Germany, Korea, Japan, and China.

Hence, the AMP Steering Committee asserts that the United States must establish a national economic framework that sets a strategy and takes supporting action to restore America's economic health and long-term strength in advanced manufacturing.

To achieve this goal, the Steering Committee recommends that the United States pursue an advanced manufacturing agenda to improve global competitiveness in the next five years. Ensuring long-term, sustainable growth requires the United States to prepare the workforce, to attract and retain skilled workers from outside its borders, and to provide incentives that encourage long-term business investments in key global growth areas.

The Steering Committee recommends the establishment of a national strategy and common agenda for advanced manufacturing to thrive in America.

III. RECOMMENDATIONS

The Advanced Manufacturing Partnership Steering Committee has developed a set of recommendations built around three pillars: enabling innovation, securing the talent pipeline, and improving the business climate. The Committee's recommendations aim to re-invent manufacturing in a way that ensures U.S. competitiveness, feeds into the innovation economy, and grows a robust domestic manufacturing base. We focus on positioning the Nation to lead the world in new disruptive advanced technologies that are changing the face of manufacturing. We believe that several key steps should be taken, but among the most critical is to strengthen our innovation system for advanced manufacturing. While some of the Nation's largest firms have the depth to be ready for the manufacturing challenges of the future, there are over 300,000 small and midsized firms that are largely outside the U.S. innovation system.[17] The United States will only lead in advanced manufacturing if it harnesses the strength of its innovation system through the manufacturing sector to create technological advantage.

The United States can and will lead the world in advanced manufacturing. Already today, we see examples of new manufacturing technologies emerging from research laboratories that will have a disruptive effect on the way things are made. Examples include novel nano--manufacturing technologies that reduce the cost of capital dramatically, bio-manufacturing and separation methods that lower the energy consumption of conventional processes, innovative additive processes and materials that reduce waste, and intelligent manufacturing tools and methods that reduce hazards, optimize supply chains, and maximize yields. Each of these innovation examples directly affects factors such as the cost of capital, quality of materials, and availability of energy.

Critical to the deployment of new advanced manufacturing technologies will be a skilled workforce trained and ready to lead this revolution in manufacturing. Exciting examples of novel partnerships between industry, educational institutions, and the public sector have come to the attention of the Steering Committee that address skills gaps in manufacturing. These partnerships are at the regional level and engage community colleges. A focus on these best practices and participation of all players (government, industry, and academia) will lead to further innovations in education and new excitement for the careers that will be created by a vibrant advanced manufacturing sector in the United States.

We see significant opportunities to exercise policy "levers" that improve the business climate for domestic manufacturing. In addition to important tax and trade policies that level the playing field, we see opportunities to engage regulatory agencies early in the development of manufacturing processes to develop a more streamlined regulatory framework and to update energy policy as well.

Finally, the Steering Committee's recommendations include concepts that can accommodate both the regional and the national aspects of any manufacturing strategy. We envision a set of regional Manufacturing Innovation Institutes (MIIs) that bridge the gap between research and commercial application of advanced manufacturing technologies. These public-private partnerships will form a national infrastructure network that eases access to new technologies while also supporting educational efforts in these new technologies. Unlocking advanced manufacturing innovation at a regional level is critical to transforming U.S. global competitiveness in manufacturing by enabling unique partnerships that leverage regional competencies. This regional focus also strengthens the collective "industrial commons" of the nation.[18]

These three pillars are closely interrelated. No one set of recommendations stands on its own. Real progress will require coordinated action with respect to all three pillars. In the following sections, the AMP Steering Committee discusses each of the three pillars and the recommendations that comprise each pillar.

Pillar I: Enabling Innovation[19]

Recommendation #1: Establish a National Advanced Manufacturing Strategy

The research and innovation ecosystem of the Nation is highly dependent on the presence of a manufacturing base that provides constant feedback in terms of problems and challenges to be solved.

Product innovation is most effective and efficient when coupled with intimate knowledge and control over the manufacturing process. Hence, the design of the product inherently involves the design of the manufacturing process by which the product will be made. The two are inseparable; severing them, as is being done increasingly often, has a very adverse effect on each because they are so interdependent.

Table 2. Summary of Recommendations

Pillar I: Enabling Innovation	
1	*Establish a National Advanced Manufacturing Strategy* Through a systematic process to identify and prioritize crosscutting technologies, a national advanced manufacturing strategy should be developed and maintained.
2	*Increase R&D Funding in Top Cross-Cutting Technologies* In addition to identifying a "starter list" of crosscutting technologies that is vital to advanced manufacturing, the AMP Steering Committee has laid out a process for evaluating technologies for R&D funding.
3	*Establish a National Network of Manufacturing Innovation Institutes* Manufacturing Innovation Institutes (MIIs) should be formed as publicprivate partnerships to foster regional ecosystems in advanced manufacturing technologies. These MIIs are one vehicle to integrate many recommendations.
4	*Empower Enhanced Industry/University Collaboration in Advanced Manufacturing Research* The treatment of tax-free bond-funded facilities at universities should be changed in order to enable greater and stronger interactions between universities and industry.
5	*Foster a More Robust Environment for Commercialization of Advanced Manufacturing Technologies* The AMP Steering Committee recommends actions to connect manufacturers to university innovation ecosystems and create a continuum of capital access from start up to scale up.
6	*Establish a National Advanced Manufacturing Portal* A searchable database of manufacturing resources should be created to serve as a key mechanism to support access by small and medium-sized enterprises to enabling infrastructure.
Pillar II: Securing Talent Pipeline	
7	*Correct Public Misconceptions About Manufacturing* Building excitement and interest in careers in manufacturing is a critical national need, and an advertising campaign should be undertaken as one important step in this direction.
8	*Tap the Talent Pool of Returning Veterans* Returning veterans possess many of the key skills needed to fill the skills gap in the manufacturing talent pipeline. The AMP Steering Committee makes specific recommendations on how to connect these veterans with manufacturing employment opportunities.
9	*Invest in Community College Level Education* The community college level of education is the "sweet spot" for impact on the skills gap in manufacturing. Investment in this sector should be increased, following the best practices of some of the leading innovators in this space.
10	*Develop Partnerships to Provide Skills Certifications and Accreditation* Portability and modularity of the credentialing process in advanced manufacturing would allow coordinated action of organizations that feed the talent pipeline.
11	*Enhance Advanced Manufacturing University Programs* Universities should bring new focus to advanced manufacturing through the development of educational modules and courses.
12	*Launch National Manufacturing Fellowships & Internships* The creation of national fellowships and internships in advanced manufacturing is recommended to bring needed resources but more importantly national recognition to manufacturing career opportunities.
Pillar III: Improving the Business Climate	
13	*Enact Tax Reform* A set of specific tax reforms should be enacted that "level the playing field" for domestic manufacturers.
14	*Streamline Regulatory Policy* A framework for smarter regulations should be created for advanced manufacturing.
15	*Improve Trade Policy* Specific trade policy proposals are advanced to improve the business climate.
16	*Update Energy Policy* Energy issues of importance in manufacturing must be addressed.

Source: AMP Steering Committee

Technology is always advancing. What was only recently on the cutting edge can quickly become a commodity. Thus, a major goal of the Advanced Manufacturing Partnership should be to develop and establish a permanent mechanism to identify the next generation of advanced manufacturing technologies that will have the greatest impact on the growth and competitiveness of the United States.

Historically, the United States has had a vibrant manufacturing base and active programs in both basic and applied research. The distinguishing feature of U.S. research activity has been the sheer scale, breadth, and vitality of U.S. investments.

Unlike the United States, other leading industrialized countries are using a more systematic planning process that is explicitly aligned to their national interests and strategies. There are benefits to implementing key elements of a structured planning process. We recognize, however, that U.S. strengths lie in flexibility and ingenuity, along with assets such as research universities and private and national labs. In cases where the risk to develop a novel, breakthrough technology is too great to be borne by one entity alone, publicprivate partnerships can accelerate the transformation of ideas to marketable goods while de-risking the investment during development. By leveraging underlying strengths that enable U.S. manufacturing enterprises to be responsive to changes in the global market, and combining them with an appropriate amount of structure, innovation in key, crosscutting manufacturing technologies will be accelerated.

The Federal Government, industry, and academia must collaborate on the creation of a sustainable process that fosters the efficient identification and commercialization of technologies that will fuel the future success of manufacturing in the United States.

To do so, we recommend that a technology lifecycle process be followed. The mechanism should have four distinct phases:

- **Phase I: Create a National Advanced Manufacturing Strategic Plan & Objectives:**
  - The AMP Steering Committee acknowledges and supports the recommendations recently published in the National Science and Technology Council's report "A National Strategic Plan for Advanced Manufacturing."[20]
  - Moving forward, we recommend that the Advanced Manufacturing National Program Office,[21] coordinate the creation of a national advanced manufacturing strategy in close collaboration with industry and academia. During this phase, future scenarios and forecasts would be created based on the analysis of strategic national (defense, energy, health, security, economic) and global (market) needs, as well as forecasted macroeconomic trends. This analysis should be conducted every five years and include industrial, academic, and government leaders and should be an inclusive process inviting opinions, using collective intelligence and building up consensus among participants. Criteria for prioritizing goals should be aligned against U.S. national security needs (defense, energy, food, health, and economic), global market demand, U.S. readiness for commercial competitiveness, and global technology readiness.
  - Table 3 lays out a framework and a directional view of the nature of the analysis required. The relative importance (high to low) and readiness assessment (high to low) will define resulting implications and define the type of technology

required to drive U.S. competitiveness. It will also guide what role the U.S. Government, industry and universities should play to advance the technology.

Phase I Output→Prioritized list of strategic needs and required technologies

- **Phase II: Create Technology Roadmaps:**
  With the national priorities in hand, working teams of industrial, academic, and agency experts should be commissioned to develop roadmaps to enable strategic planning for developing new technologies and transferring them into existing supply chains. The roadmaps should include guidance on key value and performance metrics. For mature industries this exercise should be driven by consortia composed of industrial, government, and academic leaders. For nascent technologies, the Federal government should establish working teams composed of key subject matter experts.

Phase II Output→Technology roadmaps for each of the prioritized technologies

- **Phase III: Create and Manage Programs:**
  - Based on the technology roadmaps created in Phase II, the AMP Steering Committee recommends that multiyear programs with stable funding be established to develop research capacities and create an institutional hub for coordinating technology transfer to existing firms and the retraining of incumbent workers. Wherever possible, it is critical that a co-funded model be used wherein both industry and government contribute. For mature industries, consortia should create and manage the programs. For nascent industries and technologies where the government plays a larger role in driving research and infrastructure, and is therefore the primary stakeholder, programs should be managed by dedicated program managers from Federal agencies. Wherever possible, established programs should use the proposed Manufacturing Innovation Institutes to conduct research, and develop and maintain the talent pipeline for industry. It is critical that the more than 300,000 small and medium--sized enterprises and members of the extended value chain are also involved and gain the required access to research infrastructure. Due to programs being funded by a variety of stakeholders, policies must clearly define intellectual property access rights for industry participants.
  - We recommend a competitive selection process be used for disbursing project funding based on clearly established metrics for proposal evaluation and awards. Specifically, the NIST gateoriented approach is recommended and should include such key metrics as novelty of approach, impact on tradability/ differentiation, business case, and return on investment from commercialization of the technology.

Phase III Output→Technology programs established and executed

- **Phase IV: Review Progress and Correct Course:**
  - Key stakeholders, agency representatives, and experts from academia and industry should conduct periodic reviews of programs to identify key successes and course correction needs. Standing program review teams should provide real-time technical assistance and ongoing and iterative advice as the program ramps up. While program funding must be stable, we recommend that allocations within the portfolio be subjected to review and adjustment based on rigorous, metric-based analysis—such as measure of commercialization rates, number of small and medium-sized enterprises (SMEs) served, reductions in the amount and types of energy used, or education and retraining successes. The implications of changing macroeconomic conditions should also be considered.

Phase IV Output→Periodic review of program portfolio by key stakeholders

Table 3 provides a framework for setting priorities for advanced manufacturing investments.

Recommendation #2: Increase R&D Funding in Top Cross-Cutting Technologies

Eleven crosscutting technology areas were selected as the initial list on which the Advanced Manufacturing National Program Office should focus its attention. While time did not permit the development of detailed technology roadmaps for each of these technologies, they consistently emerged as the top candidates for further consideration through consultations with key stakeholder groups, including the AMP Steering Committee itself, AMP Steering Committee Regional Meeting participants, and members of MAPI, the National Center for Manufacturing Sciences (NCMS) and the Association of Public and Landgrant Universities (APLU). (See Annex 1.)

These technologies address key national needs such as defense, energy independence and efficiency, food security, homeland security, and health care. They are pivotal in enabling U.S. manufacturing competitiveness, both in terms of differentiation and tradability of goods. Universities, national labs, intermediate technology institutes, independent research institutions, and community colleges will need to work together with industry to support research, development, and deployment of these manufacturing technologies, and to develop the talent pipeline for industry.

- Advancing Sensing, Measurement, and Process Control
- Advanced Materials Design, Synthesis, and Processing
- Visualization, Informatics, and Digital Manufacturing Technologies
- Sustainable Manufacturing
- Nanomanufacturing

Table 3. Framework for Priorities for Federal Investments in Advanced Manufacturing Technologies

US National Needs	Global Market Demand	US Manufacturing Competitiveness	Global Technology Readiness	Implication	Technology Required to Drive US Manufacturing Competitiveness	Role of US Government	Role of Industry	Role of University
High	High	High	High	Mature field. US strong global exporter.	Applied research & development to maintain leadership.	Strategic demand requires capability.	Leads research & production investment.	Conduct applied research.
High	High	High	Low	US positioned for strong global leadership. Technology not available.	Basic to applied research.	Strategic demand requires capability.	Defines road maps, develops technologies and establishes manufacturing capabilities & facilities.	Conduct basic research.
High	High	Low	High	US lags. Net importer.	Big investment required to dose gap.	Strategic demand drives establishing US manufacturing base.	Establish globally competitive manufacturing capabilities & facilities.	Breakthrough technology.
High	High	Low	Low	New field. High export potential. No global leader. New technology & infrastructure required.	Basic research.	Strategic demand drives research & infrastructure build.	Partner with universities & national labs to conduct basic & applied R&D & establish required infrastructure.	Conduct basic research.
High	Low	High	High	US specific need. Technology mature. Government road map drives infrastructure investment.	Infrastructure investment.	Strategic demand requires capability & drives future infrastructure investment.	Establish infrastructure to meet national demand.	Breakthrough technology.
High	Low	High	Low	US specific demand. Government road map drives research and infrastructure investment.	Basic to applied research.	Strategic demand sets requirements.	Establish infrastructure to meet national demand.	Conduct basic research.

US National Needs	Global Market Demand	US Manufacturing Competitiveness	Global Technology Readiness	Implication	Technology Required to Drive US Manufacturing Competitiveness	Role of US Government	Role of Industry	Role of University
High	Low	Low	High	US needs; others produce. Low global demand. US vulnerable.	Big investment required to close gap.	Strategic demand drives infrastructure build & incentives.	Only establish capability if government funds.	Breakthrough technology.
High	Low	Low	Low	US needs; no one produces; invention required.	Basic research.	Strategic demand drives research.	Establish infrastructure to demonstrate technology & meet national demand.	Breakthrough technology.
Low	High	High	High	US leads; strong exporter. Industry drives research based on global demand.	Applied research.	Incentivize exports.	Industry leads research & invests in production.	Breakthrough technology.
Low	High	High	Low	US leads; strong exporter. Industry consortium leads future road mapping.	Basic to applied research.	Incentivize exports.	Industry defines road maps, develops technologies and establishes infrastructure.	Conduct industry funded basic & applied research.
Low	High	Low	High	US not global leader. Commoditized market.	Big investment required to close gap.	Unless US vulnerable, no action required.	Only invest if breakthrough enables global competitiveness.	Breakthrough technology.
Low	High	Low	Low	New field. High export potential. No global leader. New technology & infrastructure required.	Basic research.	Incentivize exporters.	Drives research & infrastructure investment. Partners with universities to conduct basic research.	Conduct industry funded basic research.
Low	Low	XXX	XXX	No demand. Don't do it.				

Source: AMP Steering Committee.

- Flexible Electronics Manufacturing
- Biomanufacturing and Bioinformatics
- Additive Manufacturing
- Advanced Manufacturing and Testing Equipment
- Industrial Robotics
- Advanced Forming and Joining Technologies

Specific interests are as follows:

- **Advanced Sensing, Measurement, and Process Control (including Cyber-Physical Systems):** This set of technologies has applicability across almost all industry domains. These technologies are critical for enhancing tradability by way of end-to-end supply chain efficiency (e.g., low cost and pervasive sensors in plants and logistics systems, automatic control and coordination of systems-of-systems). In addition, megatrends of energy and resource efficiency, better safety, and higher quality also depend highly on advances in sensing and automatic process control. Finally, emerging technologies such as nanomanufacturing and biomanufacturing need specialized sensors and control models.
- **Advanced Materials Design, Synthesis, and Processing:** These technologies include the design and synthesis of small molecules, nanomaterials, formulated solutions, coatings, composites, and integrated components (e.g., photovoltaic devices). They entail integration of computational modeling, state-of-the-art synthesis tools (e.g., high throughput), and advanced research analytics (e.g., materials genome). Almost all the megatrends for the future—energy efficiency or alternate energy devices, new materials to counter resource shortages, next-generation consumer devices, and new paradigms in chemical safety and security—depend heavily on advanced materials. Advanced materials will fuel emerging multi-billion dollar industries.
- **Visualization, Informatics, and Digital Manufacturing Technologies:** This area entails research focused on embedded sensing, measurement and control systems for highly corrosive, high temperature processes impacting everything from chemical synthesis to lightweight materials to aircraft engines. It also includes control systems enabling manufacturing of high performance, highlycontrolled structures and devices. Finally, it entails modeling, simulation and visualization technologies that can optimize a product and its manufacturing in virtual space before actual physical production is started (therefore bypassing time-consuming and expensive physical testing and experimentation). The data generated can also potentially support conclusions regarding product warranties and product reliability.
  Examples of these technologies include integrated enterprise level smart manufacturing methodologies, e.g., moving directly from computational/digital design to chemical and materials planning, purchasing, and delivery to manufacturing of customized products and components. One aspect deals largely with manufacturing competitiveness through end-to-end supply chain efficiency—reduced manufacturing cycle time, lower worker injury and illness rates, higher process yields, higher energy efficiency, etc.—brought about by more networked

information, and the control and management of information across various entities in the value chain spanning multiple enterprises. The other aspect deals with the speed with which products are designed, manufactured, and brought to market, which will be a key differentiator.

- **Sustainable Manufacturing:** This approach aims to maximize every atom of matter and joule of energy. As a key national need, sustainable manufacturing involves technologies and systems that enable optimal raw material, energy, and resource utilization, including areas as diverse as high performance catalysis, novel separations, and new reactor and waste management systems. A major area of focus will be energy efficient manufacturing— where high energy-consuming manufacturing processes need to be substituted by lower energy-consuming alternatives. Areas such as remanufacturing (i.e., using recycled components) also need to be researched. In addition to savings in energy consumption and higher profitability, many accompanying benefits can aid the competitiveness of industry.
- **Nanomanufacturing:** Nanomaterials are forecasted to play a gamechanging role in applications ranging from high-efficiency solar cells and batteries, environmental control through nanotech-based filters, and nano-biosystem-based medical applications to next-generation electronics and computing devices. Similarly, microstructures on devices will play a key role in delivering new features or enhancing current functionality. The possibilities are limitless, but processes and quality control systems must be developed to reach the full potential of nanomanufacturing. The challenge will be to scale up and reduce costs.
- **Flexible Electronics Manufacturing:** Technologies for flexible electronics manufacturing will be major differentiators in the next generation of consumer and computing devices. Some of these devices are expected to be among the fastest growing product categories over the next decade.
- **Biomanufacturing and Bioinformatics:** Technologies to improve healthcare will require newer, more effective, and cheaper molecules. Food security is a key concern of the future, where bio-manufacturing, proteomics, and genomics will play a critical role. In addition, this technology has the inherent potential to enable energy efficiency in manufacturing. For instance, it offers room-temperature synthesis routes that can possibly replace current high-temperature processes. Innovations in the bio–nano interface such as bio-inspired manufacturing using self-assembly have the potential to simplify and scale up many complex and expensive nanomanufacturing technologies and make them economically viable.
- **Additive Manufacturing:** A growing application of manufacturing is the production of highly customized and personalized products. Additive manufacturing (e.g., three-dimensional printing) is a key technology that holds this promise. In addition, the technology has several characteristics that enable unique capabilities and features. For example, multiple materials can be processed, enabling smart components to be fabricated with embedded sensors and circuitry. Internal features can be produced that significantly enhance performance and therefore differentiate products (e.g., internal cooling channels that are optimized for thermal performance that are not possible with current manufacturing techniques). Also, materials can be processed efficiently with little waste, enhancing the sustainability of organizations that adopt additive manufacturing technologies.

- **Advanced Manufacturing and Testing Equipment:** Advanced manufacturing takes place worldwide. In those cases where it occurs outside of the United States, it is still possible for U.S. firms to maintain a significant advantage through the production and supply of high-value manufacturing equipment, such as bioreactors, CNC machine tools, or other high-technology production tools. Being the supplier of choice of advanced capital equipment will continue to yield advantages in terms of innovation and advanced engineering, as well as economic benefits.
- **Industrial Robotics:** Automation and use of industrial robots in labor-intensive manufacturing operations, such as assembly, product inspection, and testing can enable high endurance, speed, and precision. Equally important is their use in processing high temperature, corrosive and toxic substances, and materials. This technology has great potential to enhance safety and productivity of the U.S. workforce and enable the United States to compete with lowcost economies, both for domestic and export markets. Future needs in this area are being driven by the intersection of bionanotechnologies and their associated manufacturing needs.
- **Advanced Forming and Joining Technologies:** Most current mechanical manufacturing processes continue to depend largely on traditional technologies, mainly for metals, such as casting, forging, machining, and welding. These technologies will continue to be mainstays of future production processes. However, there are new and expanding needs for joining a wider variety of materials with greater energy and resource efficiency. In addition, improved performance requires continued innovation and the search for transformative technologies that will help maintain U.S. competitiveness in industries ranging from transportation to infrastructure.

Recommendation #3: Establish a National Network of Manufacturing Innovation Institutes

The United States has a long history and solid reputation for being a global leader in research and discovery. This achievement has been enabled by our preeminent research universities and national laboratories. Many of our research discoveries, however, have not been quickly translated into U.S.-manufactured products. Many technologies fail to move to commercialization because the private sector, particularly SMEs, often does not have adequate technical resources and is not able to make sufficient investments in early technologies. In fact, the stage between research and production is a perilous period in business development and is often called "the valley of death."

In part, the valley of death is attributable to the significant differences in the way activities in research and in manufacturing are conducted. Basic research and new discoveries tend to be curiosity driven, with the end goal often being validation of an idea. Conversely, manufacturing activities are competitive, focused, and systematic, driving system engineering to design, develop, and scale replicable high yield, high quality, low cost products and processes. Figure 3 depicts the gap in investment between technology readiness level (TRL) 4, the technology development phase, and TRL 6, the technology demonstration phase.

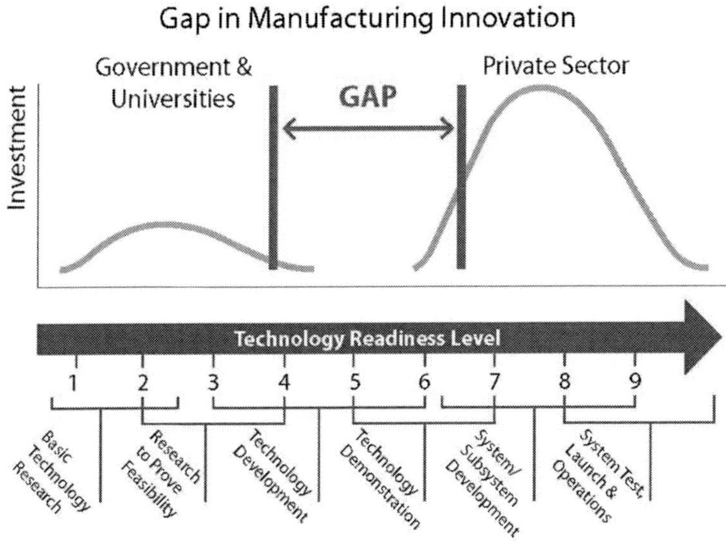

Source: AMP Steering Committee

Figure 3. Manufacturing Innovation: Investment Gap.

Extensive benchmarking (see Annex 2) identified several desirable attributes of a shared national infrastructure for supporting the translational activities for bridging fundamental research to manufacturing:

- Long-term partnership between industry and universities, enabled by Federal, State, and local governments;
- Sustained focus on technology innovation with a strong brand identity and reputation;
- Ability to identify critical emerging technologies with transformational impact and capacity in translating these technologies into products and businesses for the market;
- Ability to form effective teams of industrial and academic experts from multiple disciplines to solve difficult problems from precompetitive research to proprietary technology or product development;
- Dual appointments of faculty and students in both research universities and application-oriented institutions to develop leaders familiar with research applications, new technologies, and production systems;
- Ability to engage and assist SMEs that need new technologies by providing highly trained personnel to work in multiple regional innovation centers; and
- Ability to assist community colleges to develop and offer courses in various manufacturing technologies.

To enable the United States to successfully translate discoveries into products or applications in manufacturing, we recommend the establishment of a national network of Manufacturing Innovation Institutes (MIIs) to bridge the gap between basic research performed in universities and national laboratories, and our production enterprises, particularly SMEs. These Institutes would serve as anchors for technology development, education, and workforce training as illustrated in Figure 4. In effect, the MIIs would function as embedded nodes within a distributed network of research institutes concurrently anchoring both a national and a regional innovation system.

MIIs should support priority areas in crosscutting manufacturing technology, focusing initially on those recommended above, and subsequently on priority new technologies as they arise. Future areas of support would be expanded to include areas of emerging technologies that have the greatest potential for commercialization into new products, and adoption to create faster, cleaner, and better production processes. These areas are to be identified using the proposed model and roadmap process for prioritizing investment in advanced manufacturing technology. An open, competitive process with peer review should be used to establish the MIIs.

Each Manufacturing Innovation Institute should:

- Focus on an area of U.S. national economic strength or a promising emerging technology.
- Be hosted by an industry consortium (two or more members) and a university or national lab. New or existing partnerships can apply for government matching funds to create an MII with the membership of two or more large companies, the participation of related SMEs, and at least one major research university, with active participation by other regional universities and community colleges.
- Be governed by a Board of Directors composed of representatives from business, academic, and government organizations supporting the MII.
- Operate independently with contractual flexibility, with the provision that all MIIs will be members of the national network and will follow a similar governance model defined by a national governing board.
- Be staffed with fulltime applied researchers, engineers, and innovation enablers who support the process of technology commercialization, industrial scientists and engineers in residence, part time faculty, postdoctoral researchers, and student interns.
- Serve as hands-on "training centers" for university and community college manufacturing programs.
- Conduct projects that include precompetitive research and proprietary technology and product research, with a strong intellectual property (IP) protocol that favors manufacturers.
- Receive support via a mixed funding model from industry, academia, and government, with government (State or Federal) funding guaranteed for a minimum of 5 years with the potential of renewal for a total of 10 years.
- Receive an industrial 1:1 match to government funding for each MII.
- Establish distributed manufacturing support centers throughout the region to assist SMEs that want to adopt new technologies.
- Provide assistance to community colleges wishing to develop and strengthen advanced manufacturing programs.
- Provide grants to other universities and businesses that are developing complementary and enabling technologies.

- Provide a shared infrastructure for technology development and serve as a "collaboratory" for research universities and businesses by providing existing and startup businesses with greater access to research, students, internships, workforce development, technology transfer, and commercialization.

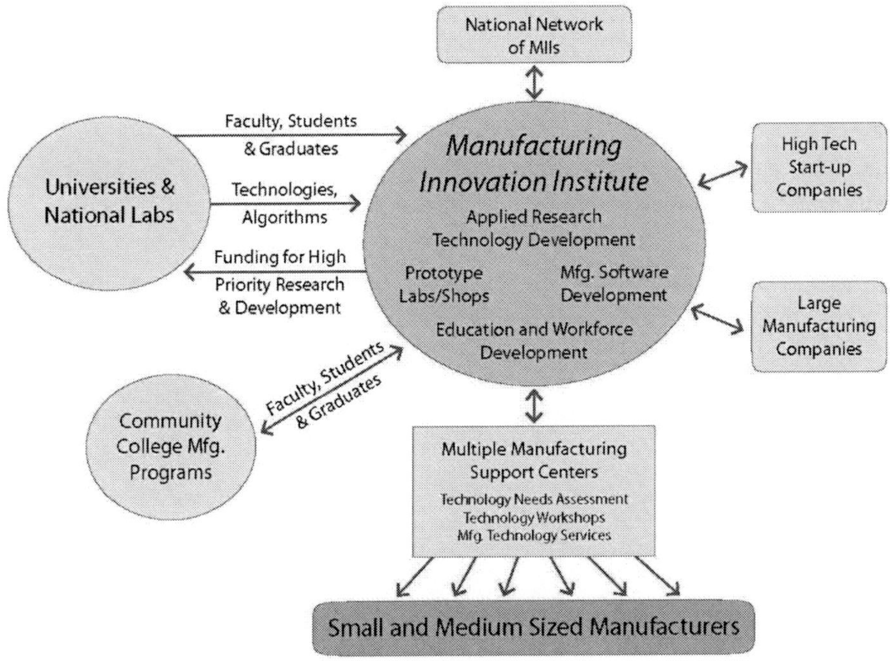

Source: AMP Steering Committee

Figure 4. Manufacturing Innovation Institute Model

- Provide a variety of business services such as design, digital manufacturing, prototype and test services, and staff training.

Recommendation #4: Empower Enhanced Industry/University Collaboration in Advanced Manufacturing Research

Achieving the full potential for the development of Manufacturing Innovation Institutes also requires a U.S. commitment to reinvigorate the environment for industry and university research collaboration. Robust research partnerships between industries and universities are a historical strength of the United States. The imperatives created by increased global competition in emerging technologies, the demands for new models of interdisciplinary research, and the shorter time horizons between fundamental discovery and applied research necessitate an examination of opportunities to enhance the climate for robust industry/university collaboration in research and commercialization.

The Advanced Manufacturing Partnership Steering Committee has identified a critical need to deepen industry and university collaboration and invest more resources at the nation's leading universities. The evolving nature of global competition creates imperatives for more rapid project and agreement development, and at times, greater focus on exclusive rights and licensing arrangements.

To remove policy barriers to more robust partnership development, we recommend the United States take action to end the restrictive tax policies that impede the speedy development of industry/university research collaborations and partnerships. Specifically, we recommend establishing a waiver mechanism or exception to Revenue Procedure 200747 to remove the cap on private-use activities in buildings constructed with tax-exempt bonds.

Recommendation #5: Foster a More Robust Environment for Commercialization of Advanced Manufacturing Technologies

There is a critical need to more fully integrate manufacturers into the robust innovation ecosystems that have evolved in universities over the last several decades. These ecosystems commonly feature seed funds, mentoring, incubation, and entrepreneurship training programs. While we recommend fostering stronger integration between manufacturers and university innovation ecosystems, we also realize that fundamental barriers impede SMEs from engaging with university research and that access to capital for new technologies is limited. A starting point to address this challenge is to create stronger synergy and a true continuum between programs to aid startups and those that are capable of supporting the scale-up of emerging manufacturing technologies through innovative, targeted procurement initiatives.

> *"Since January 2006, less than 10% of all U.S. venture capital dollars went to seed funds investing in financing rounds in the $1-4 million dollar range, and 69% of those dollars went to three states."*
>
> *—Small Business Administration, Early Stage Innovation Small Business Investment Company (SBIC) Initiative, March 2012*

Next steps could include:

- Building a manufacturing component into university innovation ecosystems
    - Incorporate manufacturing impact measures into the annual performance reports issued by the Association of University Technology Managers that reflect domestic production and employment captured from both startups and licensing activity. Including these measures would place manufacturing front and center in university technology transfer strategy development and stimulate a vibrant exchange on best practices. The annual measures would encourage a greater focus on manufacturing in regional economic development partnerships and among universities and manufacturers in the development of sponsored research partnerships.
    - Build stronger linkages between manufacturing support resources and university efforts to support startups by expanding the work of the nation's Manufacturing Extension Partnership (MEP) centers to create direct supply chain development, prototyping, and early stage engineering services for advanced manufacturing spin outs. This action will place the MEP's and manufacturing considerations at the heart of university spin out support activities.

"The Small Business Investment Company invested $2.5 billion in FY2011 in high growth small businesses. According to the program statistics, about one in five dollars between 2007 and 2010 directly supported manufacturers."

—Small Business Administration, Agency Financial Report, FY2011 and Program Statistics

- Foster a continuum of enhanced capital access from start up to scale up
  - Create a special "Phase 0" section of the Small Business Administration's Small Business Innovation Research (SBIR) program. This program would provide support for the critical preearly stage funding activities associated with testing the commercial potential of new technologies—including early prototype development and market development.
  - Expand the resources available for early stage growth and accelerate startup interaction with major manufacturers. Mechanisms such as the NSF-created 501(c)3 Innovation Accelerator should be expanded nationwide to support startups emerging from Federal advanced manufacturing research programs.
  - Clear the pathway from startup to pilot scale production by generating greater interagency coordination of procurement programs such as Defense Production Act Title III funding. The Title III program is intended to provide the Department of Defense with "a powerful tool to ensure the timely creation and availability of domestic production capabilities for technologies that have the potential for wide-ranging impact on the operational capabilities and technological superiority of U.S. defense systems." We recommend the creation of a formal collaboration between the Advanced Manufacturing National Program Office and Department of Defense Title III program, as well as other relevant Federal procurement programs, to establish a focused effort that can help complete a continuum of capital support from pre-company formation through to early phase pilot and scale up production.

Recommendation #6: Establish a National Advanced Manufacturing Portal

SMEs in the manufacturing sector are a critical component of the U.S. economy, representing 84 percent of manufacturing establishments in 2009[22] and employing 51 percent of the U.S. manufacturing workforce in 2010.[23] Their growth is inextricably linked to our continued prosperity, and they are an important source of innovation. A key driver of that growth is information—specifically, technical assistance and resources. Our work revealed that SMEs are hampered by the lack of access to this technical assistance and information. It is scattered across numerous databases and individual websites, requiring time-consuming research to access.

Firms as well as experts reported that conventional web searches for such information did not produce useful results. This problem is in part due to the vast variation and complexity of the research and innovation conducted in cooperative research centers. Simply put, finding practical answers to basic questions common to advanced manufacturing is onerous for small firms with limited time and R&D staff.

To address this issue, we propose the creation and launch of a National Advanced Manufacturing Portal: a single online destination where companies, organizations and individuals can search for federally funded cooperative research centers that best meet their needs. With this harmonized central repository of information from various sources, firms can develop short and long-term R&D plans. A National Advanced Manufacturing Portal would advance the goal of pushing innovation down the supply-chain by providing businesses with the ability to plan their process innovations as well as the design and development of new products. It would connect SMEs to the existing network of publicly-funded R&D resources that are intended—by legislative intent and design—as access points for them to gain technical assistance and information about advanced manufacturing processes.

A National Advanced Manufacturing Portal would provide an updated, harmonized catalog of information about the portfolios of the cooperative research centers and the most frequent kinds of technical assistance and resources requested by SMEs. It would also allow state and local science and technology policymakers to ascertain existing federally funded resources so they could be leveraged. In addition, it would allow researchers to determine the relative coverage of science and technology resources in a given area or targeted technology. This knowledge could lead to more efficient science and technology policy investment and coordination.

The up-front resource requirements to launch a National Advanced Manufacturing Portal are relatively small, since cooperative research centers could provide and update the information on their own facilities through a web reporting interface using a standard format. This reporting would produce the content portion of the portal. The online information clearinghouse itself would need to be hosted and maintained long-term by the appropriate Federal agency.

We propose that the initial implementation of the portal should be limited to peer--reviewed facilities such as grant recipients of public funding to ensure the quality of facilities listed.

The portal would also provide answers to frequent questions such as the availability of training, fees for access, presence of scale-up facilities on site, and whether production runs could be conducted. It would include a search by keyword feature.

The portal would be launched using the mailing lists of existing programs. For example, the NIST Manufacturing Extension Partnerships have more than 7,000 client firms.[24] We recommend that NIST serve as the administrative host agency to coordinate with portal initiatives in the Department of Energy, Department of Defense, and Federal Government agencies focused on aspects of the preproduction process. The database would be linked to websites such as Manufacturing.gov and nonprofit and public web portal networking initiatives such as Autoharvest.com.

Pillar II: Securing the Talent Pipeline25

To renew and revitalize its manufacturing prowess and competitive edge, the United States must continue to generate a steady stream of skilled manufacturing professionals. However, in recent years, persistent public misperceptions about the manufacturing sector have taken hold, tarnishing its image as a desirable long-term career focus. The false conventional wisdom about the manufacturing sector is that the work is based on repetition of tasks and evocative of the past. The main misperceptions about employment in manufacturing can be summarized by the three "D's" that characterize the work as dull, dirty, and dangerous.

In addition, with the loss and export of millions of manufacturing jobs over the past few decades, a career in manufacturing is seen as offering little, if any, job security, and no long-term career development path.

The reality is that manufacturing can be exciting, engaging, essential and environmentally sustainable, offering a pathway to upward mobility and the realization of the "American Dream."

A recent report from Booz & Company noted that a "contributing factor to this employee scarcity is traditional manufacturing's lack of appeal to students." The company surveyed more than 200 undergraduate students in engineering, science, and mathematics, and found that only 50 percent of the engineering students and 20 percent of the math and science students regarded manufacturing as an attractive career. The report also observed that "[a]round the same time, Siemens reported having nearly 3,500 open manufacturing positions" in the United States requiring STEM skills, but "with low expectations of filling many of them."[26]

As fewer students select careers in manufacturing, the demand for manufacturing-related programs across all sectors of the educational system is sharply reduced. These institutions, K12 through universities, respond by deemphasizing manufacturing-related curricula and courses in classical engineering and engineering technology programs such as two and four year mechanical and manufacturing engineering programs.

Perhaps the worst misperception of all among young people is that "America is not committed to remaining a manufacturing powerhouse in the world and that all manufacturing will eventually be done outside our borders."[27] This perception must be reversed. For the United States to remain competitive, talented employees with a high level of technical skill are needed to revitalize, sustain, and improve U.S. manufacturing.

Employment opportunities for skilled operators and technicians are increasing at a rate exceeding the availability of qualified candidates, impeding industrial growth. These innovative professionals are essential to a corporation's long-term competitiveness. In a recent survey conducted by the Manufacturing Institute and Deloitte, manufacturing was nationally viewed as core to our economic prosperity and preferred as an industry for creating local employment (Figure 5). Respondents ranked new manufacturing facilities first when asked what type of new industry facility they would support to create 1000 new jobs (Table 4).[28]

Another survey of manufacturers by the Manufacturing Institute and Deloitte on available skills to support manufacturing growth revealed that 82 percent of manufacturers reported moderate-to-serious gaps in the availability of skilled manufacturing candidates.[29] Fifty six percent anticipated the shortage to grow worse in the next three to five years.[30] Additionally, 74 percent of manufacturers reported that this skills gap has negatively impacted their company's ability to expand operations. This skills gap has resulted in five percent of all manufacturing jobs going unfilled, even in the face of our current unemployment levels.[31] To close this gap, the focus needs to be on securing and developing a strong pipeline of prepared manufacturing candidates as a key enabler to advancing manufacturing in the United States.

The Nation needs to rely upon successful public-private partnerships if it is to effect lasting improvements in its approach to education and talent development. The AMP Steering Committee looks both to the Nation's classrooms and its veterans to find the game-changing concepts that will secure the manufacturing talent pipeline going forward.

Table 4. Ranking by respondents of the type of new industry facility they would support to create 1,000 new jobs in their communities

Facility	Rank
Manufacturing facility	1
Energy production facility	2
Healthcare facility	3
Technology development center	4
Communications hub	5
Retail center	6
Financial institution	7

Source: Deloitte and the Manufacturing Institute, "Unwavering Commitment: The Public's View of the Manufacturing Industry Today," 2011 Annual Index, themanufacturinginstitute.org/ ~/ media/ 2AB778520C734D888156A90B667C1E70.ashx.

Challenges facing the manufacturing industry cannot be successfully addressed by individual companies, academia, or Federal agencies alone. Instead, the opportunities must be examined through partnerships to identify and use the best solutions. Partnerships are a strong platform from which to address the quickly changing needs of manufacturers in the United States. Identifying and responding to technological innovation can be expensive and time--consuming, but these hurdles are best overcome through partnerships of similarly motivated groups. Developing a highly skilled professional manufacturing workforce is as critically important to the 21st Century manufacturing sector as any other single element. We studied examples of successful partnerships to identify and review best practices and attributes. While no two partnerships are alike, even in the same region with similar partners, successful workforce partnerships of academia, industry, and government share the following six characteristics. See *Annex 3* for a full list of the resources and exemplary partnerships that formed the foundation for identification of these characteristics.

PROJECT SHINE (SHAPING HIGH-QUALITY INTEGRATED NEBRASKA EDUCATION)

Created in 2009, Project SHINE's goal is to increase student interest and participation in high demand technical careers. Project SHINE integrates Nebraska's manufacturing, energy, biofuels and food processing businesses with secondary and college STEM educators and their students.

Project SHINE's approach to engaging education and business professionals in teaching and learning is by exposing educators and students to "realworld" business environments, and build partnerships between education and business. Each group benefits through opportunities to participate:

- Educators participate in externships with business partners through curriculum development workshops that focus on problembased learning.
- Students explore careers through the Nebraska Career Connections and summer camp activities that focus on how math and science are used for business applications.

- Industry partners benefit through the growth of potential pipeline of local skilled technicians for their business.

Since 2009, 72 STEM educators and almost 5,000 Nebraska middle and high school students have participated in the National Science Foundation-funded Project SHINE. An additional 200 students participated in the SHINE STEM summer camps. Project SHINE's impact will reach even further as nearly 200 Project Based Learning (PBL) resources developed through the project are disseminated through an online e-library.

Source: Project SHINE Webpage, mechatronicsmec.org.

- **Have a Passion for Learning and a Vision for the Future:** Successful manufacturing partnerships have at their roots a shared passion for lifelong learning. They have a vision of success and commitment to being globally competitive. Benchmarking, research, and exploring are essential ingredients of the most advanced partnerships. There is a balance between incumbent worker training and new worker training for young people coming into the workforce from the K12 and community and technical school systems.
- **Embrace Change:** The future of manufacturing will be radically different from its past. The status quo curricula, teaching methods, and silos must be replaced with a collaborative, innovative, lifelong learning culture. A nation rich with human capital simply cannot tolerate the loss of its competitive edge and international standing in the science, technology, engineering, and mathematics (STEM) disciplines.
- **Convene Organizations to Share Their Expertise:** In order to facilitate and drive change, each region or community will need a respected organization capable of convening the necessary parties. The most effective convening organizations are neutral non-profits, knowledgeable about, but not responsive to, political or other influences. Professional organizations and academic institutions are well positioned to act as convening organizations. The Manufacturing Innovation Institutes could serve this critical convening function and link education and training efforts to firm networks already engaged in the R&D services provided by these institutes. By using the MIIs as convening spaces for both technology development and workforce development programming, there is an institutional recognition of the synergy between building a regional labor market and sustaining innovation capacity.
- **Collaborate with a Common Purpose:** To be effective, manufacturing partnerships must be based on mutual respect and a spirit of cooperation, a candid recognition of strengths and weaknesses, and shared goals reached in the spirit of collaboration and consensus. The academic community offers the expertise to teach students, but learning must be developed and applied by collaborating with industry, which in turn, needs a steady stream of STEM talent, and must become vastly more engaged in K20 education and community activities. Community colleges should deliver the curricula associated with the technical training of both new and incumbent workers. The implementation would involve industry and technology experts recommending industry-specific skills in order for these curricula to stay current.

- **Have Clear, Specific Roles:** In order to align partners towards a common goal, the partners should negotiate and agree first upon the goal, then upon the resources and roles that each of the stakeholders will bring to the partnership.
- **Maintain Flexibility:** Manufacturing technology is developing as rapidly as information technology. To adapt to this increasing pace of change, all parties must be ready and willing to adapt and act quickly. This adaptation will require flexibility and collaboration among all parties, and the coordinated use of existing training and educational facilities to prevent wasted time and redundancy.

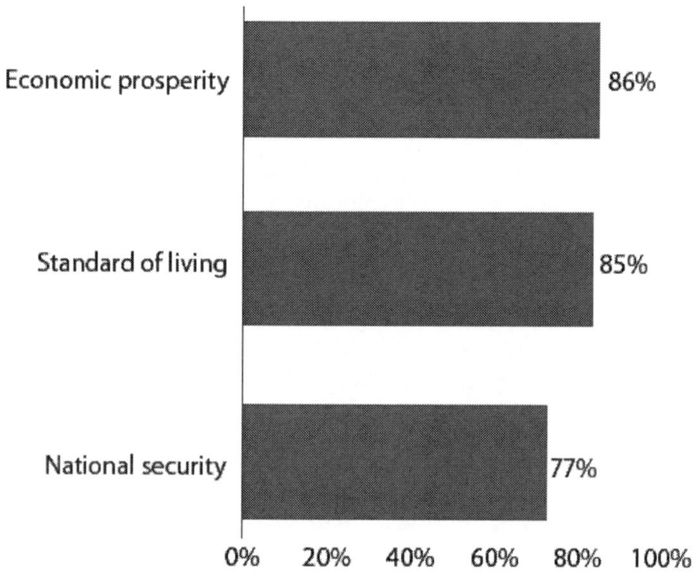

Source: Deloitte and the Manufacturing Institute, "Unwavering Commitment: The Public's View of the Manufacturing Industry Today," 2011 Annual Index, themanufacturinginstitute.org/~/ media/ 2AB778520C734D888156A90B667C1E70.ashx.

Figure 5. Percentage of respondents who believe the manufacturing industry is very important to U.S. economic prosperity, standard of living, or national security.

Effective manufacturing partnerships build a sustainable culture of dynamic change so that acceleration in manufacturing technology is injected back into education and training. Change introduced through partnerships is most effective when it meets the fundamental needs of people and society and is presented consistently and clearly. Project-based learning provided in partnership with manufacturing ensures that the subject matter is relevant and applicable to developments outside the classroom. The U.S. Government's role is most effective when it funds impactful academic programs; defunds ineffective programs; sets the boundaries to ensure fair opportunity to all; and fills the gaps that industry and academia are unable to fill.

As education provides a foundation for the future, there is a need to modify traditional teaching methods used to train the manufacturing workforce at all levels. Success in advanced manufacturing and entrepreneurship will require fundamental STEM skills as well as broad problem-solving skills, decision-making skills, and people skills that do not emerge from a conventional K20 education. It is necessary to pursue initiatives that can expand the manufacturing workforce in the near-term, such as support for veterans programs and community colleges, which can expand the future pipeline.

The benefits of manufacturers participating in academia at all levels are many for the students who choose careers in manufacturing. Effective manufacturing partners engage K--12, community colleges, undergraduate and graduate students and professors through internships, apprenticeships, projects, research, co-ops, endowments, scholarships, and hiring. These offer opportunities for academics and students alike to stay current in the field of manufacturing and contribute to research and discovery.

Academia should bring manufacturing directly into the classroom, providing real world projects and research opportunities that engage students and inherently provide professional development to teachers. Teachers across our K20 system must be encouraged to embrace project-based learning of all varieties, take advantage of summer internships, and engage with manufacturing experts. In turn, industry simply must be an active, responsive partner in the classroom.

The AMP Steering Committee encourages the adoption of project-based learning of all kinds, and to varying extent based on the local needs, across the K20 spectrum; a number of these projects should be selected for their relevance to manufacturing-relevant skills, such as supply-chain management, design for manufacturability, estimation of tolerances and requirements, economics, and team management. To stimulate these new initiatives, educational partnerships between industry, academia, and local and regional government must be established.

Recommendation #7: Correct Public Misconceptions about Manufacturing

In order to lay the foundation for a secure, sustainable manufacturing talent pipeline, the AMP Steering Committee recommends the creation of an aggressive, integrated "Image of Manufacturing" public service announcement campaign that would raise awareness and correct misperceptions about manufacturing in the United States.

Recommendation #8: Tap the Talent Pool of Returning Veterans

Veterans have many of the work-life and job skills that are in high demand and that are often missing in the general workforce. These include maturity, discipline, and the ability to work effectively in groups and leadership. In addition, many veterans have undertaken extensive technical training, resulting in skills that could be easily transferred to manufacturing positions, and have become technicians, operators of complex equipment, and craftsmen. Despite their skills, the veteran population that served in the military in the period since September 2001 (called Gulf Warera II veterans) is experiencing a higher rate of unemployment than its civilian counterpart. The Bureau of Labor Statistics found the unemployment rate for this cohort of veterans was 12.1 percent in 2011.[32] Similarly, with the withdrawal of troops from Iraq and Afghanistan and potential cuts to Defense Department funding, the number of veterans seeking employment is certain to rise in the coming years. There are two obstacles to overcome in bringing veterans to manufacturing careers: low awareness of the opportunities in the sector and difficulty equating military skills with private sector job qualifications. We recommend providing a training module on the career opportunities in advanced manufacturing to the Department of Defense's Transition Assistance Program, which provides support and information to transitioning veterans about career options post-service. In addition, the Departments of Defense and Labor should accelerate their efforts to categorize military occupational codes and translate them to civilian skills, as well as providing the opportunity for active duty servicepersons to earn professional accreditations.

Recommendation #9: Invest in Community College Level Education

Manufacturing jobs have changed, requiring highly skilled workers. The largest gap between manufacturing employers' needs and new employee skills exists in the occupations of technician and equipment operator. This gap has left many workers unqualified for available positions. Community colleges currently provide training for the missing skills, but a significant gap remains.

Community colleges already enroll many of the people who should train for advanced manufacturing careers, and these institutions have partnerships, infrastructure, and teaching methods that are focused on regional needs. They grew after World War II to train returning GIs to join the workforce. This founding principle can help train and retrain today's workforce to meet the needs of local manufacturers. Investing in community colleges and promoting engagement between community colleges and industry, universities, national labs, and K12 programs are important steps. Modest changes to governmentfunded grant opportunities to encourage partnering with community colleges could create stronger regional partnerships with industry, universities and national laboratories.

To advance these programs, the AMP Steering Committee recommends:

- Modification of the Department of Education's Graduate Assistance in Areas of National Need (GAANN) program to have a focused solicitation on manufacturing fellowships/scholarships at the university and community college level. This opportunity could be structured to encourage collaboration between industry, community colleges, and universities or have separate scholarship programs aimed at the different educational levels.
- Creation of a national network of manufacturing educators by integrating educational programs among the National Science Foundation, the Department of Education, and the Department of Labor in order to share best practices, curricula, and resources. This national network could build on the institutional infrastructure provided by the national network of Manufacturing Innovation Institutes. Industry and national manufacturing associations and societies should be included in the network.
- Proposal and implementation of changes to align ongoing solicitations for federally funded research programs to encourage partnerships with community colleges. The MIIs should serve as a resource for trained personnel who can assist community colleges in developing appropriate courses as well as providing hands-on projects and coordination of internships with regional manufacturing companies.

Recommendation#10: Develop Partnerships to Provide Skills Certifications and Accreditation

The AMP Steering Committee recommends a national focus on education and training that can produce workers capable of operating and troubleshooting modern factory equipment. The approximately 1,500 community colleges located across the United States provide an opportunity to develop location-specific curricula to meet the needs of local manufacturers. We encourage the accreditation of programs, the development and standardization of community college curricula where appropriate, and the use of stackable professional credentials to meet the needs of manufacturing.

In order to support the development of a robust and high-skilled pipeline of manufacturing talent, national standards, credentials, and certifications are critical to provide manufacturing the consistent baseline ability to qualify candidates as to their educational, behavioral and leadership knowledge, manufacturing experience, and individual competencies. An efficient market for employees with necessary knowledge and skills depends on reliable and appropriate credentials and certifications. To succeed, any new credentials and certifications require a critical mass of national recognition, and acceptance and adoption by industry, education, and government. Such credentials and certifications work when they:

- Involve quality assessments, accurately gauging worker skills;
- Include an accreditation regimen that ensures the quality of the program and alignment with the changing needs of industry;
- Evolve and continuously update to accommodate the changing needs of workers and the manufacturing sector; and
- Result in nationally portable, industry-recognized credentials that support preferential consideration and job search mobility.

This recommendation can be implemented by leveraging existing efforts. In 2009, the Manufacturing Institute partnered with ACT, Manufacturing Skills Standards Council, National Institute for Metalworking Skills, American Welding Society, and Society of Manufacturing Engineers to build a comprehensive model for certifications, which can provide the core framework and partnerships. This partnership is called the Manufacturing Skills Certification System (MSCS), and it is in the process of adding more industry-recognized credentials at this time.

Aligned with this partnership, we recommend support for a coalition of industry associations, professional societies, and educational organizations to establish a national framework of standards, accreditations, and certifications at each level of the advanced manufacturing competency model. (See Figure 6.)

> *Right Skills Now is a partnership between government and industry that is building a fast-track, 16-week training program that supports the workforce needs of manufacturers of all sizes.*
>
> *-Manufacturing Institute, Right Skills Now*

We further recommend that an accreditation-like review system be set up regionally for community college manufacturing programs, and that the requirements for professional credentials be used as a foundation for curricular development in high schools, community colleges, and universities.

Any path forward must incorporate two key functions—accreditation and certification—and result in two key outcomes—common education and training standards—to satisfy current and emerging competencies, and yield portable certifications for individuals.

National associations involved in manufacturing should initiate and coordinate a register of certifications that are available in all regions and can be "stacked" one after another to assemble complete programs of training in advanced manufacturing.

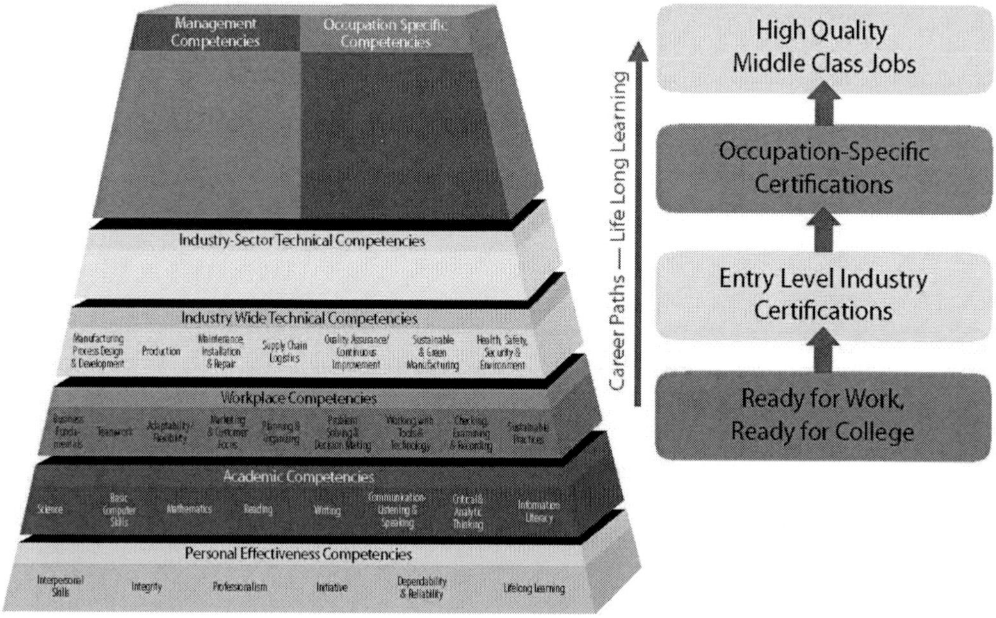

Source: Advanced Manufacturing Competency Model, Adapted from Department of Labor's Employment and Training Administration http://www.careeronestop.org/ competencymodel/ pyramid.aspx?hg=Ywww.careeronestop.org/competencymodel/pyramid.aspx?hg=Y.

Figure 6. Department of Labor Advanced Manufacturing Competency Model.

Recommendations #11 and #12: Enhance Advanced Manufacturing University Programs, and Launch Manufacturing Fellowships and Internships

Major research universities in the U.S. have a key role in defining the fundamental elements of advanced manufacturing and developing the next generation of educators and industrial leaders. They can qualitatively advance the manufacturing profession and retool its image as a challenging and rewarding career.

Universities, however, are uncertain about where the discipline of manufacturing best fits in academia. It does not fit well into normal boundaries of degree programs, departments or even schools, and as a result often finds itself marginalized.

The AMP Steering Committee therefore recommends augmenting existing engineering curricula with manufacturing coursework, and creating new graduate-level programs that provide students with a comprehensive overview of manufacturing as well as technological and operational perspectives in a professional engineering context.

Public-private partnerships are critical to ensuring U.S. manufacturing excellence to implement this recommendation. As with community college programs, it is expected that the Manufacturing Innovation Institutes will play a significant role in providing course development and hands-on training, as well as assisting with internships in local manufacturing establishments. This recommendation requires implementation at multiple levels:

- **University level:** Establish new masters-level professional degrees in manufacturing leadership at research universities. These degree programs should be comprehensive in their integration of technologies, such as robotics and advanced automation, with methods, such as supply chain management and human systems integration.
- **Government level:** Fund National Manufacturing Fellowships and Veterans Leadership in Manufacturing Fellowships, traineeships, and curriculum development. These initiatives could help to correct the misperceptions of manufacturing held by the future workforce.
- **Industry level:** Encourage industry participation by providing a professional career path for graduates of advanced manufacturing programs. By teaming with universities on both curricular and career development, industry can help to solidify the profession of manufacturing and make it a highly attractive field for study and practice. Internship programs can expose students to careers in manufacturing leadership. During the educational process, industry representatives should serve as mentors and role models to students entering and emerging from these new programs and provide relevant, hands-on manufacturing experiences for graduate students.

While it is impossible to separate the educational system into discrete pieces, we believe the most impactful recommendations are at the high school and community college levels, followed by undergraduate education. Industry has identified deficiencies in both content mastery and soft skills. These skills could be taught through the use of project-based learning, which has proven its unique capacity to prepare young people for a 21st century workplace. The deficiencies in content mastery and soft skills can be remedied through this approach when coupled with a marked improvement in teaching skills.

Pillar III: Improving the Business Climate33

The United States is at risk of losing leadership in manufacturing, most importantly its ability to manufacture the high-technology products that are invented and innovated in this country. To attract investment and production, the United States must promote a competitive business environment, which includes a robust talent pipeline, 21st century infrastructure, and strong investment in R&D. While these are important to the overall health of the U.S. economy, they are particularly important for the advanced manufacturing sector, which faces intense global competition. Major economic competitors are making it increasingly attractive for companies to invest in locations outside the United States.

The AMP Steering Committee does not believe that it is the role of government to formulate a national industrial policy of direct investment in or subsidies to specific firms. However, we recommend that the United States create a national framework to create a favorable business climate for manufacturing that spurs investments and fosters partnerships across government, academia and industry.

For the United States to continue to be an attractive location for businesses, we recommend building a policy framework that spurs investments and fosters partnerships between government, academia, and industry. The foundation of that framework should be constructed through targeted policies in four areas:

- Tax Policy
- Regulatory Policy
- Trade Policy
- Energy Policy

Recommendation #13: Enact Tax Reform

A key focus of the Advanced Manufacturing Partnership is on the linkage between U.S.-based innovation, R&D, and manufacturing. To encourage investment in the United States, we recommend that the corporate tax system create a more attractive environment for businesses to compete globally. The United States has the highest statutory corporate tax rate, including Federal and State taxes, among the 34 members of the Organization for Economic Cooperation and Development. This system is an impediment to U.S.-headquartered businesses and businesses interested in investing in the United States.

Comprehensive U.S. tax reform is particularly important for the advanced manufacturing sector. Manufacturing is a source of direct and indirect highpaying jobs. Our current tax system discourages domestic capital investment in manufacturing, thereby undercutting the stability of the innovation and jobs engine that has produced unparalleled economic prosperity in the last century. The tax system distorts investment by industry, with manufacturing, construction and other high-wage and asset intensive industries paying a globally noncompetitive statutory tax rate. The result is a decrease in aggregate investment in manufacturing.

For these reasons, there is a need to reform the tax system to address the existing distortions and disincentives for manufacturing in the United States. A more favorable tax climate would serve a twofold benefit: provide incentives for U.S.-based businesses to increase investment and encourage more foreign direct investment in the United States, leading to an increase in investment, innovation, and jobs. Tax reform should also yield a tax system that is internationally competitive with others around the world in attracting and retaining advanced manufacturing and its associated innovation engine.

While there is a need for broad tax reform to make all U.S. companies more internationally competitive, our recommendations are targeted to the promotion of advanced manufacturing in the United States. They add up to an integrated package of proposals that address the mobile nature of capital and intellectual property, and enhance the incentive for retaining and reinvigorating the historical strength of closely connected U.S. R&D and production capabilities. We recommend that additional tax incentives should flow to those entities that engage in all three critical advanced manufacturing roles: U.S.-based innovation, R&D, and manufacturing.

We propose lowering the corporate tax rate to bring it more in line with other advanced economies. A rate reduction, combined with a broadening of the tax base, would encourage additional investment in manufacturing by U.S. corporations, and would position the United States as a more attractive region for direct investment by foreign corporations.

We recommend the following actions:

- *Recognize the importance of manufacturing through the tax code.* Given global competition and the ripple effect of manufacturing to the economy, any tax reform should encourage investment in manufacturing. This can be done through a reduction of the tax rate for domestic manufacturing activity.
- *Strengthen R&D tax credits.* Increase the R&D alternative simplified credit to 20% and make it permanent.
- *Create an internationally competitive corporate tax system.* The U.S. tax system must be redesigned in a way that encourages companies to invest in the United States by addressing the current law on foreign earnings of U.S.-based companies. In addition to lowering the overall corporate rate, reform must consider the tax treatment of overseas earnings of U.S. based corporations, including the consideration of a competitive partial exemption system similar to the type adopted recently by the U.K. or a minimum tax regime like Japan. Ultimately, comprehensive tax reform must ensure that U.S. companies are competitive when operating abroad and in the United States.

We recognize that efforts to address long term U.S. fiscal issues may well bring about significant proposals that include a mix of rate reductions. We urge that this debate be particularly mindful of the imperative to recognize how such actions may impact the climate for advanced manufacturing. We also caution against any new measures that could impede an improved climate for U.S.-based production or discourage investment in the United States.

Recommendation #14: Streamline Regulatory Policy

Regulation is an oft criticized, but vital function carried out by government. Well-conceived, science-based, and effectively implemented regulations are important tools for protecting consumers, workers, and the environment. When done right, regulation provides important benefits for society and can encourage greater competence and confidence in industry. Done excessively or inappropriately, or without adequate attention to its consequences, regulation can hamper innovation and international competitiveness. We recommend the following:

- **Early Engagement:** Collaboration between regulators and the regulated community can drive significant improvements in the quality of final rules. Robust dialogue between agencies and businesses ideally should occur well before the comment period. Improved use of the Advanced Notice Rulemaking Process[34] would allow manufacturers to contribute to cost-benefit analyses in a meaningful way that could make compliance more cost-effective.
- **Objective Cost Benefit Analyses:** We recommend that cost benefit analyses and risk assessments rely on the best available science.

Recommendation #15: Improve Trade Policy

A fair and open international trading system provides the greatest opportunities for U.S.-based innovative manufacturing and, ultimately, for sustaining current and creating new jobs.

We recommend that the U.S. Government take the lead on a progressive trade policy, building on recent successes such as passing the U.S.-Colombia, U.S.-Korea and U.S.-Panama Free Trade Agreements (FTAs). FTAs level the playing field for U.S. exporters, eliminating tariff barriers to market access, reducing nontariff barriers, and allowing access to dispute settlement systems.

Trade policy is an important consideration for manufacturers choosing to site new facilities, and the United States must not let its competitors outpace it in the race to negotiate further agreements. The Nation must prioritize policies that help ensure access to foreign markets and promote global competitiveness; these policies must include a focus on nontariff barriers and export control policies. The Trans-Pacific Partnership (TPP) is an example of a high-standard, groundbreaking negotiation that will cover new emerging barriers for cutting-edge technologies, promote regulatory coherence, address competition with state-owned enterprises, and provide a template for economic integration across the Asia-Pacific region.

In balance with trade liberalization, the U.S. Government should ensure a strong focus on enforcing trade rights, particularly addressing market-distorting subsidies, unfair trade practices, and intellectual property violations to level the playing field for U.S.-based manufacturing.

As near-term goals, the United States should:

- **Pursue increased market access:** The future key barriers are not tariffs; they are nontariff barriers—regulatory and standards impediments that represent *de facto* market barriers. Examples of non-tariff barrier areas are innovation principles, regulatory reform and customs facilitation, forced technology transfer and weak intellectual property enforcement. We recommend that the U.S. Government strengthen the interagency process to create a consistent agenda on regulatory issues and to strengthen cooperative, capacity-building initiatives with other key trading partners.
- **Launch new negotiations:** The U.S. Government has actively solicited input from industry on core economic trading partners for new negotiations. A number of regions, such as the Middle East and North Africa, can benefit from near-term capacity-building efforts that could lead eventually to full trade liberalization efforts. In the interim, we recommend that the U.S. Government prioritize a TransAtlantic Partnership (TAP) negotiation that would leverage the advanced economies of the United States and the European Union and allow both partners to address trade barriers as a model for future multilateral trade liberalization.
- **Reform export controls:** We recommend that the U.S. Government accelerate the reform of outdated export control regimes. This process could start with rebuilding the U.S. Munitions List by harmonizing export control licensing and administrative procedures across all involved agencies and transitioning them to a single information technology platform.

Recommendation #16: Update Energy Policy

Energy is a basic building block for today's advanced manufacturing applications. Advanced manufacturing uses innovative technologies to add value to inputs in order to produce modern materials and solutions, including electronic materials, pharmaceutical breakthroughs, and clean energy alternatives. However, energy policy in the United States must fully account for the impacts of energy costs to manufacturers and the potential to drive investment into new markets and applications as the United States seeks to transition to a sustainable energy future. Therefore, any effort to reinvigorate advanced manufacturing in the United States would not be complete without an examination of energy policy that seeks ample supplies to catalyze economic growth and prosperity. We recommend the following steps:

- **Focus on energy efficiency and conservation:** Energy efficiency is the most affordable and most available way to lower energy costs and reduce carbon emissions and is particularly important to the manufacturing sector. Every dollar saved through energy efficiency efforts can be redeployed to expand business and preserve manufacturing jobs. For example, according to the Brookings Institute, if all eligible buildings in the United States were retrofitted over the next decade, it would create roughly 215,000 direct jobs, 127,000 of which are in manufacturing.[35] We recommend policies that provide incentives for power generators and distributors to undertake cost-effective and innovative energy efficiency measures and the promotion of tools and incentives to assist manufacturers of all sizes in implementing energy efficiency measures.
- **Increase and diversify domestic supplies:** Economic growth will continue to rely on hydrocarbon energy, whether from oil, naphtha, natural gas, ethane, or coal, and will require additional domestic supplies to improve energy security and reduce price volatility. These inputs are critical for the manufacturing process as both fuel and feedstock, serving as the basic building blocks of materials used in 96 percent of all manufactured goods, including products enabling the further development of renewable sources of energy such as solar panels and wind turbine blades. Onshore, increased supply from unconventional sources, such as natural gas, oil and natural gas liquids from shale will be important resources for the United States over the next several decades. The availability of these resources for value-added products must be a policy imperative to ensure economic growth and job creation. Producers and regulators need to work together to ensure that potential reserves can be brought to market in an environmentally acceptable manner at an affordable cost. Natural gas at stable, competitive prices will continue to incentivize U.S. manufacturers to invest and create jobs in the United States. Today, industrial uses of natural gas as a feedstock are driving multibillion dollar investments. In turn, multiplier effects from these investments will be felt across the economy, including by other U.S. manufacturers that are less dependent on hydrocarbon feedstocks.
- **Speed development of renewable sources of energy:** There is a role for government, industry, and academia to work together to accelerate the development of effective and more sustainable alternative energy sources, including renewable sources. As global demand for clean sources of energy grows, the United States has the opportunity to play a key role in the manufacturing of advanced technologies such as energy storage equipment, photovoltaics, and wind power technologies. Since 2008, the United States has nearly doubled renewable energy generation. In 2011, U.S. solar installations grew 109 percent with the overall solar market surpassing $8.4 billion.[36] However, renewables remain a small fraction of U.S. energy use. Policies are needed that primarily focus on driving down costs, which

will help drive increased demand. We recommend the continued extension of financial incentives for public/private research into promising technologies and storage devices. Further, any incentives that spur the early adoption of innovative technologies such as low and no-carbon sources originating from coal, solar, natural gas, wind, tidal, and geothermal energy must be targeted at technologies that demonstrate a path towards economically viability.

- **Transition to a low carbon economy:** We believe that to create a sustainable energy future over the long term the United States needs to shift to a low carbon economy. The right mix of fundamental research, innovation, and aggressive implementation is needed to achieve both this transition and continued economic growth. The development and implementation of a broad portfolio of technologies is essential for this transition. The United States has the technical capacity to accelerate development of sustainable energy options, but large-scale commercialization of new capital-intensive manufacturing solutions will require increased public-private partnership. We recommend a targeted approach to promote aggressive basic R&D with accelerated demonstration and deployment of clean energy and next generation energy efficient technologies. Government policy can help most in specific situations, such as when the costs and market development risks of critical technologies exceed the commercial capabilities of individual companies, where the regulatory or liability risks are beyond the capacity of the private sector, and when investment timelines exceed the private sector's capabilities.

IV. CONCLUSION

The Advanced Manufacturing Partnership Steering Committee offers a comprehensive set of recommendations built around three critical pillars:

- Enabling innovation
- Securing the talent pipeline
- Improving the business climate

These recommendations are aimed at reinventing manufacturing in a way that ensures U.S. competitiveness, feeds into the Nation's innovation economy, and invigorates the domestic manufacturing base. Rather than debate whether the manufacturing jobs lost in past decades can return, we should instead focus on leading the world in the new technologies that are changing the face of manufacturing. We stress the vital importance of strengthening the U.S. innovation system for advanced manufacturing.

With sustained focus, alignment of interests, and coordinated action to implement the recommendations outlined in this report, the United States can and will lead the world in advanced manufacturing. Already, we see examples of new manufacturing technologies emerging from research laboratories that will have a transformative effect on the way America makes things.

Industry, academia, and government, nationally and locally, must act now to ignite the ingenuity to make it in America.

Together the Nation must commit to reinvent the manufacturing base to ensure its future.

APPENDIX A. ACKNOWLEDGMENTS

The members of PCAST and the AMP Steering Committee acknowledge the assistance of the following individuals during the course of this investigation:

- Alan Anderson, IDA Science and Technology Policy Institute
- Asha Balakrishnan, IDA Science and Technology Policy Institute
- Thomas Kalil, Office of Science and Technology Policy
- David Katz, National Economic Council
- Sridhar Kota, Office of Science and Technology Policy
- Thomas Kurfess, Office of Science and Technology Policy
- David Lindley, IDA Science and Technology Policy Institute
- Stephen Moilanen, National Economic Council
- Justin Scott, IDA Science and Technology Policy Institute
- Stephanie Shipp, IDA Science and Technology Policy Institute

APPENDIX B. EXPERTS CONSULTED

The members of PCAST and the AMP Steering Committee appreciate the contributions of the following individuals who were consulted during the course of this investigation. Many other experts contributed ideas through the AMP Steering Committee's regional meetings.

- Pasquale Abruzzese, Honeywell
- Ryan Adesnik, Stanford University
- Joe Asiala, MITECH+ and Blue Water Angels
- Gretchen Baier, The Dow Chemical Company
- Naleesh Bam, Honeywell
- Russell Barton, U.S. National Science Foundation
- Abby Benson, Massachusetts Institute of Technology
- Suzanne Berger, Massachusetts Institute of Technology
- William Bonvillian, Massachusetts Institute of Technology
- Elaine Brock, University of Michigan
- Bruce Brown, Procter and Gamble
- Larry Burns, University of Michigan
- Gardner Carrick, Manufacturing
- Scott Cooper, Procter and Gamble
- David Cote, Honeywell
- Richard Cowan, Georgia Institute of Technology
- Stephen Cross, Georgia Institute of Technology
- Lauren Culver, U.S. Department of Energy
- Lynn Daniels, U.S. Department of Energy
- David Danielson, U.S. Department of Energy
- Joe DeSarla, Honeywell
- Jim Davis, University of California, Los Angeles
- Michael Dombrowski, Johnson & Johnson
- David Dornfeld, University of California, Berkeley

Institute
- Christopher Cerone, Zimmer Corporation
- Don Chaffin, University of Michigan
- Fu-Kuo Chang, Stanford University
- Frank Chong, U.S. Department of Education
- Leo Christodoulou, U.S. Department of Energy
- Jennifer Clark, Georgia Institute of Technology
- Peter Cleveland, Intel
- John Fraser, Florida State University
- Erica Fuchs, Carnegie Mellon University
- Ken Gabriel, U.S. Department of Defense, Defense Advanced Research Projects Agency
- Patrick Gallagher, U.S. Department of Commerce, National Institute of Standards and Technology
- Ram Ganapathy, Honeywell
- John Garone, Honeywell
- Marc Giroux, Corning
- Jason Gorey, U.S. Department of Defense
- Nancey Green Leigh, Georgia Institute of Technology
- Jeff Hamner, Procter and Gamble
- Andy Hannah, Plextronics
- David E. Hardt, Massachusetts Institute of Technology
- Pat Healey, Procter and Gamble
- Susan Helper, Case Western Reserve University
- Gregory Henschel, U.S. Department of Education
- Gary Herrigel, University of Chicago
- Byron Hill, Honeywell
- David S. Hoiriis, Honeywell
- Carrie Houtman, The Dow Chemical Company
- Jack S. Hu, University of Michigan
- Chris Downing, Georgia Institute of Technology
- David Drabkin, Northrop Grumman
- Johnny Dwiggins, Employers in Support of the Guard and Reserve (ESGR)
- Karen Elzey, Skills for America's Future
- Kate Emmanuel, Ad Council
- Joseph Ensor, Northrop Grumman
- Jim Evans, Stryker
- Stephen Ezell, Information Technology and Innovation Foundation
- Gary Fedder, Carnegie Mellon University
- Jerry Jasinowski, independent consultant
- Don Johnson, Omni Tech International
- Henry Kelly, U.S. Department of Energy
- Thomas Kenny, Stanford University
- Dale King, U.S. Department of Education
- John Klein, Honeywell
- Robert Knotts, Georgia Institute of Technology
- Kevin Kolevar, The Dow Chemical Company
- Theresa Kotanchek, The Dow Chemical Company
- Art Kracke, Allegheny Technologies Incorporated
- Bruce Kramer, U.S. National Science Foundation
- Sanjay Krishnan, Honeywell
- Brian Krzanich, Intel
- Richard Lester, Massachusetts Institute of Technology
- Cam Mackay, MAPI
- Mike Mayberry, Intel
- Duncan McBride, U.S. National Science Foundation
- Don McCabe, Corning
- John McIver, Procter and Gamble
- Steve McKnight, U.S. National Science Foundation
- Tim McNulty, Carnegie Mellon University

- Karen Huber, Caterpillar Inc.
- Catherine Hunt, The Dow Chemical Company
- Pamela Hurt, Society of Manufacturing Engineers
- Krishna Mikklineni, Honeywell Michael Molnar, U.S. Department of Commerce, National Institute for Standards and Technology
- Siddhartha Niyogi, Honeywell
- Charles O'Hara, Procter and Gamble
- Neal Orringer, U.S. Department of Defense
- Burak Ozdoganlar, Carnegie Mellon University
- Panos Papdopolous, University of California, Berkeley
- Tom Peterson, U.S. National Science Foundation
- Gary Pisano, Harvard University
- Kameshwar Poolla, University of California, Berkeley
- Douglas R. Pratt, Genoa Associates
- G. Ranganath, Honeywell
- Timothy Regan, Corning
- Iris Rivero, Texas Tech University
- David Rosen, Georgia Institute of Technology
- Ed Rozynski, Stryker
- Gerhard Salinger, U.S. National Science Foundation
- Tariq Samad, Honeywell
- Al Sanders, Honeywell
- S. Shankar Sastry, University of California, Berkeley
- Martin Schmidt, Massachusetts Institute of Technology
- Al Schwabenbauer, independent consultant
- Paul Wright, University of California, Berkeley
- Euisik Yoon, University of Michigan
- Sam Yoon, U.S. Department of Labor
- Michael McQuade, United Technologies Corporation
- Shreyes Melkote, Georgia Institute of Technology
- Sridhar Seetharaman, Carnegie Mellon University
- Douglas Seymour, Osram Sylvania
- Susan Shields, University of Michigan
- Albert Shih, University of Michigan
- Stan Sidor, South Piedmont Community College
- Phillip Singerman, U.S. Department of Commerce, National Institute of Standards and Technology
- Costas Spanos, University of California, Berkeley
- Karen Stang, Northrop Grumman
- William Swanson, Honeywell
- Dan Swinney, Manufacturing Renaissance Council
- Rebecca Taylor, National Center for Manufacturing Sciences
- Debbie Tekavec, Carnegie Mellon University
- David Touretzky, Carnegie Mellon University
- Tana Utley, Caterpillar, Inc.
- Shankar Venugopal, Honeywell
- Kelly Visconti, U.S. Department of Energy
- Ben Wang, Georgia Institute of Technology
- John Wassick, The Dow Chemical Company
- Albert J. Wavering, U.S. Department of Commerce, National Institute of Standards and Technology
- Josh Whitford, Columbia University
- H. S. Philip Wong, Stanford University
- Jan Youtie, Georgia Institute of Technology

End Notes

[1] President's Council of Advisors on Science and Technology, "Report to the President on Ensuring America's Leadership in Advanced Manufacturing," June 2011, www. whitehouse.gov/sites/ default/files/microsites/ ostp/pcast-advanced-manufacturing-june2011.pdf.

[2] *Ibid.*

[3] *Ibid.*

[4] Stephen J. Ezell and Robert D. Atkinson, *The Case for a National Manufacturing Strategy,* Information Technology and Innovation Foundation, 2011, www2.itif.org/2011-national-manufacturing-strategy.pdf.

[5] Census Bureau, Statistics of U.S. Businesses, 2009, www.census.gov/econ/susb/.

[6] Bureau of Labor Statistics, Current Employment Statistics, 2010, bls.gov/ces/cessizeclass.htm# TB_inline? height=200&width=325&inlineId=ces_program_links.

[7] Ross DeVol et al., "Manufacturing 2.0: A More Prosperous California," Milken Institute, June 2009, www.milkeninstitute.org/pdf/CAManufacturing_ES.pdf

[8] Bureau of Economic Analysis, Industry-by-Industry Total Requirements after Redefinitions (Producer Price Indexes), 2010, www.bea.gov/iTable/index_industry.cfm.

[9] The Manufacturing Institute, *The Facts about Modern Manufacturing*, 8th Edition, 2009, www.nist.gov/ mep/upload/FINAL_NAM_REPORT_PAGES.pdf. Data are presented in Figure 1.

[10] National Science Foundation, National Center for Science and Engineering Statistics (NSF/NCSES). InfoBrief NSF 12-309, March 2012, www.nsf.gov/statistics/infbrief/nsf 12309/nsf12309.pdf.

[11] National Science and Technology Council. *A National Strategic Plan for Advanced Manufacturing,* 2012. www.whitehouse.gov/sites/default/files/microsites/ostp/iam_ advancedmanufacturing_strategicplan_2012.pdf.

[12] U.S. Census Bureau, Foreign Trade Statistics, Advanced Technology Products, 2011, www. census.gov/foreign-trade/balance/c0007.html.

[13] Department of Commerce, "The Competitiveness and Innovative Capacity of the United States," January 2012, www.commerce.gov/sites/default/files/documents/2012/january/ competes_010511_0.pdf.

[14] Gregory Tassey, *The Technology Imperative* (Northampton, MA: Edward Elgar, 2007).

[15] Erica R. H. Fuchs, "The Impact of Manufacturing Offshore on Technology Development Paths in the Automotive and Optoelectronics Industries," Massachusetts Institute of Technology, June 2006, esd.mit.edu/people/ dissertations/fuchs.pdf.

[16] Michael Greenstone, Richard Hornbeck, and Enrico Moretti, "Identifying Agglomeration Spillovers: Evidence from Winners and Losers of Large Plant Openings," April 2010, emlab.berkeley.edu/~moretti/mdp2.pdf.

[17] BLS, Current Employment Statistics, *op. cit.*

[18] Gary P. Pisano and Willy C. Shih, "Restoring American Competitiveness," *Harvard Business Review 87* (July–August 2009), hbr.org/hbr-main/resources/pdfs/comm/fmglobal/restoring-american-competitiveness.pdf.

[19] Further details related to the recommendations within this Pillar can be found in Annexes 1, 2, and 4.

[20] National Science and Technology Council, "A National Strategic Plan for Advanced Manufacturing," www.whitehouse.gov/sites/default/files/microsites/ostp/iam_advanced manufacturing_strategicplan_2012.pdf.

[21] National Institute of Standards and Technology (NIST), "National Program Office for the Advanced Manufacturing Partnership Established at NIST," Press Release, December 19, 2011, www.nist.gov/public_ affairs/releases/npo121911.cfm.

[22] Census Bureau, Statistics of U.S. Businesses, 2009, www.census.gov/econ/susb/.

[23] Bureau of Labor Statistics, Current Employment Statistics, 2010, bls.gov/ces/cessizeclass. htm#TB_inline? height=200&width=325&inlineId=ces_program_links. (*STPI Calculations)*

[24] NIST, "Reexamining the Manufacturing Extension Partnership Business Model," October 2010: 7, www.nist.gov/mep/upload/MEP_Bus_Model_Full_Report_October2010_a.pdf.

[25] Further details related to the recommendations within this Pillar can be found in Annex 3.

[26] Arvind Kaushal, Thomas Mayor, and Patricia Riedl, "Manufacturing's WakeUp Call," *Strategy & Business* 64, Booz & Company and Tauber Institute for Global Operations, University of Michigan (August 3, 2011): 38, www.tauber.umich.edu/docs/ManufWakeUp_w_Cover.pdf.

[27] Joe Anderson, Council Chairman, and Mike Laszkiewicz, Workforce Development Subcommittee Chair to Department of Commerce Secretary Gary Locke, "The Manufacturing Council," July 2011, www.trade.gov/manufacturingcouncil/documents/ MC_Workforce_08222011.pdf.

[28] Deloitte and the Manufacturing Institute, "Unwavering Commitment: The Public's View of the Manufacturing Industry Today," 2011 Annual Index, www.themanufacturinginstitute.org/ ~/media/2AB778520C734 D888156A90B667C1E70.ashx.

[29] Deloitte and the Manufacturing Institute, "Boiling Point? The Skills Gap in U.S. Manufacturing," 6, www.themanufacturinginstitute.org/~/media/ A07730B2A798437D 98501E798C2E13AA.ashx.

[30] Deloitte and the Manufacturing Institute, "Unwavering Commitment: The Public's View of the Manufacturing Industry Today," 2011 Annual Index, www.themanufacturinginstitute.org/~/ media/2AB778520C734 D888156A90B667C1E70.ashx.

[31] Ibid.

[32] Bureau of Labor Statistics, BLS Economic News Release: Employment Situation of Veteranc2011, March 20, 2012, www.bls.gov/news.release/vet.nr0.htm.

[33] Further details related to the recommendations within this Pillar can be found in Annex 4.

[34] Advanced rulemaking is intended to solicit comments and information from all segments of the public interested in a particular issue prior to an agency determining whether a rule (regulation) will be proposed.

[35] Susan Helper, Timothy Krueger, and Howard Wial, "Why Does Manufacturing Matter? Which Manufacturing Matters? A Policy Framework," Brookings Institute, February 2012. www.brookings.edu/~/media/research/files/papers/2012/2/22%20manufacturing%20helper%20krueger%20wial/0222_manufacturing_helper_krueger_wial.pdf.

[36] Solar Energy Industries Association, *U.S. Solar Market Insight 2011*, March 14, 2012. www. greentechmedia.com/research/ussmi/.

In: Advanced Manufacturing in America
Editors: Brandon Cullerne and Michael Laston
ISBN: 978-1-62257-932-7

Chapter 3

CAPTURING A DOMESTIC COMPETITIVE ADVANTAGE IN ADVANCED MANUFACTURING. ANNEX 1: TECHNOLOGY DEVELOPMENT WORKSTREAM REPORT *

President's Council of Advisors on Science and Technology

EXECUTIVE SUMMARY

Background

The Technology Development Workstream had two broad objectives. The first was to determine a sustaining mechanism to be used for identifying and developing key manufacturing technologies, and the second was to set forth a set of top technology areas that would ensure U.S. manufacturing competitiveness.

To meet the objectives, the mechanisms used to gather data were mainly the following:

- Surveys to elicit opinions and feedback from industry and university participants regarding important industries for the future, and key technologies that will be needed, as well as private-public partnership best practice examples and models.
- Workshops with industry and university participants to identify the top technologies required for manufacturing competitiveness.
- Desk research of comparable mechanisms used in other countries and regions to identify and nurture technologies.
- Whitepaper solicitations from select experts.

The research and innovation ecosystem of a nation is highly dependent on the presence of a manufacturing base that provides constant feedback in terms of problems and challenges to

* This is an edited, reformatted and augmented version of the Technology Development Workstream Report, dated July 2012.

be solved. Also, product innovation cannot exist without intimate knowledge and control over the manufacturing process. In other words, design of the product also inherently involves design of the manufacturing process by which the product will be made. Historically, the United States has had a vibrant manufacturing base and active programs in both basic and applied research. The distinguishing feature of U.S. research activity has been the sheer breadth and vitality at various levels.

A review of mechanisms used in other countries indicate a use of more hierarchical, systematic planning systems that are explicitly aligned to national strategies– often necessitated by their lack of resources, as opposed to in the United States, where much more bottom-up innovation flourishes and much larger bets are made with respect to research. But in 2011, the world shifted. For the first time in history, total R&D spending in Asia Pacific exceeded total R&D spending in the U.S., with the largest increases occurring in China from government sources. This trend is forecasted to continue at even a higher rate. The use of a structured planning process at a national level has the advantages of both aligning and allocating national resources more efficiently into U.S. efforts to revitalize planning, as well as creating a platform to better address competing national strategies from other countries (that are often government led and therefore difficult for U.S. industry participants to counter by themselves). Therefore, we have incorporated this kind of a hierarchical planning process into our recommendations.

Findings

- A number of model programs have excellent best-practices that can be adopted.
- Greater industry participation is required to identify and nurture technologies—whether in determining the national strategy, developing technology roadmaps, or creating and managing programs. Industry-university-government advisory boards and consortia can work well in this regard.
- An analysis of mechanisms used in other countries and regions suggests that key differences exist in the use of national foresight mechanisms, as well as higher involvement of industry and university stakeholders.
- A broad list of crosscutting technologies that are critical for establishing U.S. manufacturing competitiveness has emerged. This list can help kick-start the sustaining mechanism for identifying and developing advanced manufacturing technologies.

High-Level Recommendations—Sustaining Mechanism for Identifying and Developing Technologies

A technology life-cycle process is proposed, using the following main steps:

- **Use a national foresight mechanism to generate a national strategy on the set of important future needs and broad technology areas of focus.** Incorporate inclusive and multiple methods to gather as much input as possible from various industry and university stakeholders and subsequently arrive at a consensus.

- **Use industry-government-university consortia (where possible) to generate detailed roadmaps for industries of strategic importance and based on the broad technology areas laid out in the national strategy.** These roadmaps should include guidance on key performance metrics. For example, if sustainability is an important objective, then there should be a common understanding of what is included in a set of sustainability metrics and how they should be measured. For the cases where industry maturity is low, the federal government has to drive the research and infrastructure development agenda. In fact, many mature industries with a strong U.S. manufacturing base are also under competitive threat, and roadmaps and investment to keep these "crown jewels" profitable are also needed.
- **Create and manage programs to carry out the research and eventually help commercialize technologies.** Industries need to have strong participation in terms of funding, setting program goals and carrying out the project-management activities. Programs need to have long-term goals (essentially run over a few years) and have stable sources of funding. Rigorous, metric-oriented analysis and review of proposals (to award projects) and performance have to be carried out.
- **Conduct periodic review of program results and strategic analyses of the portfolio of projects.** Adjustments need to be made to funding allocations in the portfolio. Consortia also need to periodically reevaluate their missions and reset their goals depending on changing external conditions.

Successful development and commercialization of advanced manufacturing technologies require that several needs be met, namely:

- The need for effective programs in universities for generating a pipeline of engineers who are engaged in applied research and early stage commercialization, and go on to be future manufacturing and R&D leaders.
- The need for policy support in the area of R&D tax incentives as well as trade incentives to ensure export competitiveness.
- The need for shared infrastructure accessible to industry members, including subject-matter experts (SMEs), for their research and development needs and in enabling commercializing of technologies.

Top Technology Areas of Focus—Cross-cutting Technologies Selected as a Starter List for Advanced Manufacturing Partnership (AMP) Focus

- Advanced sensing, measurement, and process control.
- Advanced material design and synthesis, including nanomaterials, metamaterials, metals, coatings, ceramics.
- Information technologies, including visualization and digital manufacturing.
- Sustainable manufacturing.

- Nano-manufacturing (includes micro feature manufacturing).
- Flexible electronics.
- Bio-manufacturing and bioinformatics, including proteomics and genomics.
- Additive manufacturing.
- Advanced manufacturing equipment (including testing).
- Industrial robotics.
- Advanced forming (including near net shape manufacturing) and joining/bonding technologies.

CHARGE TO THE WORKSTREAM

The Technology Development workstream was given two main objectives. The ***first objective*** was to define the best permanent mechanism for identifying and nurturing the set of technologies that will have the greatest impact on the retention and future growth of manufacturing in the United States, enabling differentiation and competitiveness for U.S. manufacturing from an end-to-end supply-chain perspective over a sustained period of time. The ***second objective*** was to recommend the top manufacturing technology areas from an industry and university perspective and to analyze some of the supporting/enabling factors and constraints that affect U.S. manufacturing competitiveness.

PROCESS FOLLOWED

The following actions were taken to define the sustaining mechanism:

- A qualitative survey was carried out with key industry personnel who have extensive experience working in joint programs with Federal agencies, national labs, and universities. They were asked for their experiences on best run programs, opinions on best practices for program management, and views on best practice partnership models from other countries and regions.
- Research of public-private partnerships and technology-development mechanisms in other countries and regions was carried out to determine alternative models.
- An analysis of the key factors that should be taken into account to identify and develop technologies was carried out.
- A white paper was solicited from a joint industry-university expert panel to analyze and lay out recommendations on improving public-private partnerships.

The results from the actions above were synthesized into defining an ideal technology lifecycle process, finding gaps in current practice, and developing specific recommendations accordingly.

To establish the list of key advanced manufacturing technologies, three separate surveys were carried out. The first was a pilot survey with an initial set of industry and university participants. The aim was to identify the manufacturing industries that will have the greatest economic and national impact on the United States now and into the next decade and then identify the set of leading advanced manufacturing technologies required. A set of interim findings was created. Feedback was elicited from broader group of universities (APLU), manufacturing members (MAPI), and small and medium enterprises (NCMS) in terms of the interim findings. Based on the results of the previous steps, a more comprehensive survey was then carried out with a broader set of industry and university respondents on the key technology areas for the future.

The team then compared these technology areas with the FY11 and FY12 Federal agency advanced manufacturing programs to verify alignment to needs and to identify technologies that would benefit from stronger public-private partnerships. Interactive workshops were held in universities to further gather input on key technology challenges for manufacturing, as well as the top technologies on which stakeholders should focus.

A high-level analysis was carried out of some key megatrends and drivers for manufacturing to determine whether the technologies selected had sufficient coverage, and a final list of top technology areas to enable U.S. manufacturing competitiveness in terms of enhancing tradability and differentiation was then synthesized from the outputs from the previous steps. This formed the starter list that could be used in the strategic-planning process.

KEY FINDINGS

Sustaining Mechanism for Developing Technologies

The survey of U.S. industry participants on public-private-partnerships elicited the following qualitative inputs:

- The best run programs have the requisite amounts of applied research and business focus and are thus often led by industry-government-university consortia.
  - The semiconductor industry is an excellent case study of consortia running applied and basic research programs over a long period of time with fairly good historic results. SEMATECH and SRC have been cited as good role model consortiums with very well run programs. Roadmapping has also been a highlight of these consortia, with the ITRS being a good example.
  - Consortia have a goal of driving pre-competitive technologies that can benefit all companies involved to move the industry forward.
- There are also excellent examples of outstanding government-led programs—for example, the National Institute of Standards and Technology Advanced Technology Program (NIST-ATP), Defense Advanced Research Projects Agency (DARPA), various Department of Energy – Energy Efficiency and Renewable Energy (DOE-EERE) programs, Department of Defense (DOD) ManTech program—that have been cited by respondents.
- Stakeholder participation and transparency are key. Industry partners need to be involved in the process of determining programs goals and strategies.

- Programs need to have long-term goals and stable funding commitments. All respondents noted that the best programs were always multiyear programs with clear funding commitments.
  - Because of the longer term view, partnerships need to periodically examine their missions and goals and make periodic adjustments.
  - A longer term perspective does not imply a purely theoretical effort; programs need to have a strong focus on (ultimately) reducing research to industrial practice.
- Partnerships require industry participants to co-fund initiatives. This brings the necessary focus on applied research and drives a results-oriented approach due to strong return-on-investment (ROI)-driven incentive.
- The prevailing view is that the government should not pick product technologies for the future (i.e., try to make big bets on specific technologies unilaterally), but rather define the problem statement fairly tightly and let the partnership work out the technologies. An example would be the Nanoelectronics Research Initiative (NRI), which has as its mission, "Demonstrate novel computing devices capable of replacing the CMOS FET as a logic switch in the 2020 timeframe."
- Partnerships need sufficient longevity and breadth to establish and pursue overall objectives that extend across multiple Federal funding agencies while remaining responsive to individual agency objectives.
- Universities are, and need to be, a key focal point for research and education mechanisms. This has worked well for some of the best run programs and, more important, is a key mechanism through which a pipeline of research and leadership talent for the manufacturing industry has to be built. An important point to note here is that federally funded programs have played a key role in supporting industry internships/graduate fellowships and generating this pipeline (e.g., the Air Force Research in Aero Propulsion Technology [AFRAPT] fellowship program).
- From a specific program-management perspective, the following are some of the key requirements:
  - Program management should involve the industry/university participants. Applied-research programs should in fact be run by program managers from industry as they have high stakes involved (i.e., their own funding and ROI).
  - The need for unbiased technology oversight committees to double-check the technologies selected in the roadmap.
  - Rigorous, metric-based project selection/reviews and portfolio management.
  - NIST-ATP's gate-based proposal practice, which begins with concept papers, followed by technical proposals and finally business proposals, has often been cited as a best practice, well-defined, gated proposal process. Early feedback on concepts helps companies unfamiliar with the process to quickly learn it and be more efficient with resources
  - Contracting or award agreements should be simplified and flexible. A key requirement is to ensure that intellectual property (IP) availability is easy for the industry partners.

The comparative study of other country/region mechanisms for identifying and developing technologies uncovered the following key findings:

- Germany, Japan and Korea are examples of high-cost economies that have focused on, and successfully managed to retain, their manufacturing bases. The following are some of the key points related to the processes and policies utilized by these countries:
    - All have clear national objectives and strategies for retaining manufacturing excellence. They use hierarchical planning mechanisms incorporating technology-foresight processes. The foresight processes themselves are highly inclusive, inviting participation from numerous industry and university experts, whether through white-paper solicitations, Delphi forecasts, brainstorming workshops, or other mechanisms. The German model also makes extensive use of data mining and bibliometrics.
    - Manufacturing still retains its prestige as a career path in these countries. Students carry out applied research in universities and institutes and go on to become manufacturing leaders in industry.
    - Government supports student internships, enabling critical hands-on learning
    - In Germany, universities and institutes are deeply involved in solving industry problems. There is a large component of applied research that is carried out in the various educational institutions in the ecosystem. German universities acknowledge and reward applied-research contributions to a greater extent than U.S. universities.
    - In Japan, applied research is driven more by government and industry consortia than universities, though that has been changing over the last decade.
    - In these countries, SMEs are considered an important part of the ecosystem, and research programs and infrastructure exist to specifically benefit them.
    - They have retained a strong grip over capital-intensive equipment globally. In other words, although a lot of the commoditized manufacturing operations have moved to low-cost economies, the high-cost capital equipment (utilizing advanced manufacturing technologies and quality) needed in the factories are often sourced from these countries.
- The European Union's 7th Framework Program (FP7) is an inclusive mechanism for identifying and developing technologies. The goals of the program are to achieve growth, employment creation, industrial competitiveness, and technological leadership. Extensive community consultation through events, workshops, and conferences is employed in finalizing the research areas of focus.
- In the semiconductor industry outside the United States, a number of consortia of government and industry have developed technologies and built industrial competitiveness. Among these are SELETE in Japan, IMEC in Belgium, LETI in France, and ITRI in Taiwan.

Advanced Technology Areas

The top industries identified through the industry survey were the following:

1. Advanced materials **(75%)**.
2. Agriculture chemicals, biotechnology, chemicals, and communication equipment **(50%)**.
3. Aerospace/defense, automotive, appliances, building materials, health care, consumer products, renewable energy **(38%)**.

The industry respondents identified the following technologies as most critical to retaining and growing a manufacturing base in the United States:

1. Advanced sensing and measurement technologies **(88%)**.
2. Sustainable manufacturing **(75%)**.
3. Process control and nano-scale materials **(60%)**.
4. Nano-manufacturing, lightweight materials, information technology, flexible electronics, coatings, and continuous process control **(50%)**.

Additional insights gained from the in-depth industry interviews also revealed the following industrial priorities and interests on potential game-changing technologies:

- **Materials genome.** Industry believes this is a game changer, particularly for nano-scale materials. A number of the advanced materials companies are already investing in this area; however, all agree that further improvements in materials modeling and optimization, scientific data sharing, and information management are required to enable the full potential of in-silico materials design and accelerated delivery of next-generation materials and processes.
- **Lightweight materials.** Industry is heavily investing in advanced materials and composite processing technologies to enable development of lightweight batteries, building materials, auto components, etc. Most programs are aligned to energy drivers—enhanced energy efficiency, energy storage, and energy generation.
- **Nano-manufacturing.** Breakthrough advances are required to consistently and economically manufacture game-changing nano-materials and chemicals. Advances in this area are tightly coupled with materials genome, advanced sensing and measurement, visualization, and process control.
- **Information technology.** Every respondent noted the critical importance of information technology and data management. It is the foundation for enabling advanced manufacturing from product and process design, to adaptive-process control- to supply- chain management.
- **Adaptive design and processes.** Industries are interested in enhanced methods for rapid design and modularization of flexible, adaptive products and processes.

The university respondents deemed the following technologies as critical:

1. Nano-manufacturing (**60%**).
2. Advanced sensing and measurement technologies (**60%**).
3. Information technology (**50%**).

The respondents also identified additional early-stage technology areas that have the potential to be future game changers:

- **Advanced material design capabilities (material genome, integrated computational materials engineering).** Advanced materials will play a huge role in future products, and developing the ability to discover and manufacture them efficiently is critical.
- **Additive manufacturing.** The ability to inexpensively produce highly customized or personalized products is the ultimate goal of manufacturing technology and will be a true disrupter. A related, but more highly developed, set of concepts would be those from DARPA's AVM (Adaptive Vehicle Make) portfolio, which aims at "a bitstream-programmable manufacturing facility that can be rapidly configured to produce a new design or design variation with nearly zero learning curve. We call this large-scale manufacturing in quantities of one."
- **Bio-manufacturing.** Because bio-products will likely play a significant role in the future, it is imperative that the right foundations are laid in this area to drive viable economics.
- **Industrial robots.** These have the potential to increase the productivity of the workforce.

A systematic analysis of the various factors that influence the evolution of manufacturing technology from the perspectives of differentiation and tradability identified six key factors:

1. Energy efficiency
2. Sustainability/green
3. Productivity
4. New technology
5. Globalization
6. Customization/personalization

A study of the megatrends that are likely to play a big role over the next decade or two—energy, health care, food security, resources management (water and minerals), safety and security, smart world—overlaid on the factors enabling differentiation and tradability, leads to a broader list of manufacturing technologies that also includes the early-stage and future game- changing technologies that emerged from the industry and university surveys:

- Bio-manufacturing.
- Additive manufacturing.
- Micro-manufacturing.

- Advanced manufacturing equipment.
- Industrial robotics.
- Advanced forming (including near-net-shape manufacturing) and joining/bonding technologies.

Combining the information gathered above, we get a final list of technologies:

- Advanced sensing, measurement, and process control (also known as smart manufacturing or advanced automation).
- Advanced material design and synthesis, including nano-materials, meta-materials, metals, coatings, ceramics.
- Information technologies, including visualization and digital manufacturing.
- Sustainable manufacturing.
- Nano-manufacturing (includes micro feature manufacturing).
- Flexible electronics.
- Bio-manufacturing and bioinformatics, including proteomics and genomics.
- Additive manufacturing.
- Advanced manufacturing equipment (including testing).
- Industrial robotics.
- Advanced forming (including near-net-shape manufacturing) and joining/bonding technologies.

This cross-cutting set of technologies has wide applicability across many industries. These technologies will also play a critical role in addressing national strategic needs (such as defense and food security). Note that there is considerable interplay between these technologies, and their effects and benefits can really be realized when they are applied in an integrated fashion.

Additional Considerations

The survey responses also uncovered other significant factors that drive investment decisions on whether to manufacture within or outside the United States. It is essential that improved policies are set up in terms of tax incentives, investment support, and trade policies to make the United States a competitive manufacturing destination.

The following are the top factors that contribute to the decision to manufacture outside the United States:

1. Proximity to customers **(90%)**.
2. Cost **(50%)**.
3. Regulations **(40%)**.
4. Tax policies **(40%)**.

The following are the top factors that contribute to the decision to bring back manufacturing to the United States:

1. Cost **(50%)**.
2. Regulations **(50%)**.
3. Proximity to customers **(40%).**
4. Tax policies **(40%).**
5. Investment required **(40%).**

Based on the responses, the biggest reason for manufacturing outside the United States is proximity to customers. This reflects the growing importance of markets outside the United States, especially in the emerging economies. This is why technologies for optimization of the overall global supply chain need to be included in our focus on manufacturing. For the decision to manufacture in the United States, the end-to-end supply-chain costs, as well as financial and regulatory factors, play a critical role. We have to ensure that we improve both:

- Differentiation, that is, have better products than competition.
- Tradability, that is, the ability to manufacture in the United States and export anywhere competitively.

In addition, two more factors are critical for successful development of advanced manufacturing technologies:

- A pipeline of trained engineers carrying out research in universities, who then go on to become manufacturing leaders.
- A network of shared infrastructure for carrying out the research needed by industry

RECOMMENDATIONS

Sustaining Mechanism for Technology Development

The sustaining mechanism is modeled as a process in the figure below.

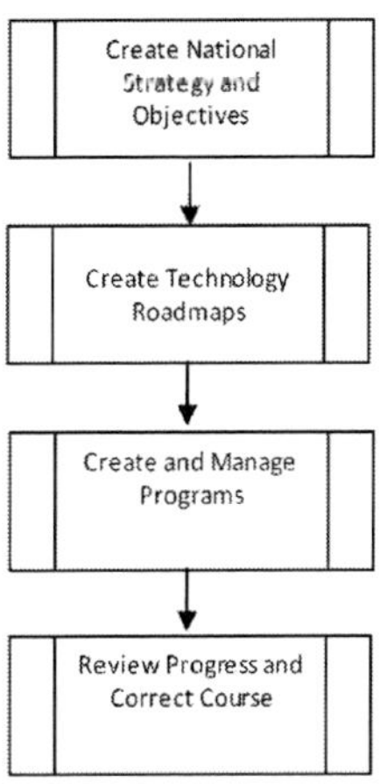

The specific recommendations are as follows:

1. The Advanced Manufacturing National Program Office should establish and coordinate the first step of creating the national strategy and objectives, that is, to set up a national foresight process. This activity needs to be completed over the next 3 months or so. The foresight process itself will be an infrequent task done once every 4 or 5 years. The next steps of the technology life cycle are highly dependent on the results of the first step, so the subsequent owners and time lines have to be determined after the completion of step 1.
2. This step of creating the national strategy and objectives needs to analyze strategic national and global needs, identify macro trends that will likely play out, and create future scenarios and forecasts. The table below lays out a guideline framework and a directional view of the nature of the analysis required. Strategic choices will be outlined in terms of the key factors (i.e., industry maturity, national need [defense, energy, infrastructure, food, economic security], global demand and technology maturity) and specific industries. Estimating the specific categorization of industries or technologies (e.g., in terms of high-, low-, or medium-technology readiness or global market demand) is not a straightforward exercise, but qualitative estimations should be carried out. The "Don't do it" scenarios below are those where public-private-partnerships are not advised (note that this is a broad, directional view of how the strategic analysis and decision-making may proceed).
   The final output will be a document that lays out the strategic national needs (and associated industries), future scenarios, and broad technology areas of strategic interest.
3. Multiple mechanisms are needed to make the national foresight process step as inclusive as possible and to drive consensus. Among these are expert panels involving industry/academic fellows, white-paper solicitations from research institutes/consortia, Delphi-style forecasts, brainstorming workshops, surveys, etc.

The sequence of steps proposed is given below:

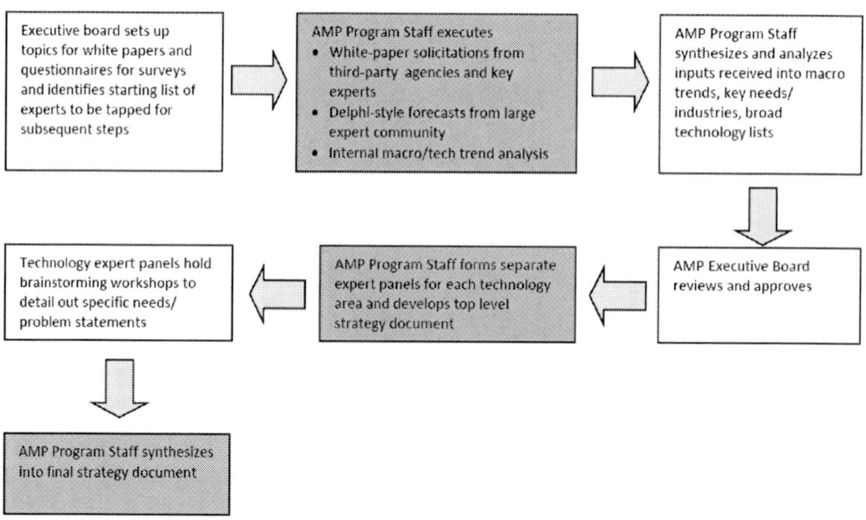

Framework for Priorities for Federal investments in Advanced Manufacturing Technologies

US National Needs	Global Market Demand	US Manufacturing Competitiveness	Global Technology Readiness	Implication	Technology Required to Drive US Manufacturing Competitivenss	Role of US Government	Role of Industry	Role of University
High	High	High	High	Mature field. US strong global exporter.	Applied research & development to maintain leadership	Strategic demand requires capability.	Leads research & production investment	Conduct applied research
High	High	High	Low	US positioned for strong global leadership. Technology not available.	Basic to applied research	Strategic demand requires capability.	Defines roadmaps, develops technologies and establishes manufacturing capabilities & facilities.	Conduct basic research
High	High	Low	High	US lags. Net importer.	Big investment required to close gap.	Strategic demand drives establishing US manufacturing base.	Establish globally competitive manufacturing capabilities & facilities.	Breakthrough technology
High	High	Low	Low	New field. High export potential. No global Leader. New technology & infrastructure required.	Basic research	Strategic demand drives research & infrastructure build.	Partner with universities & national labs to conduct basic & applied R&D& establish required infrastructure.	Conduct basic research
High	Low	High	High	US specific need. Technology mature. Government roadmap drives infrastructure investment.	Infrastructure investment	Strategic demand requires capability& drives future infrastructure investment.	Establish infrastructure to meet national demand	Breakthrough technology
High	Low	High	Low	US specific demand. Government roadmap drives research and infrastructure investment.	Basic to applied research	Strategic demand sets requirements.	Establish infrastructure to meet national demand.	Conduct basic research

Framework for Priorities for Federal investments in Advanced Manufacturing Technologies (Continued)

US National Needs	Global Market Demand	US Manufacturing Competitiveness	Global Technology Readiness	Implication	Technology Required to Drive US Manufacturing Competitivenss	Role of US Government	Role of Industry	Role of University
High	Low	Low	High	US needs; others produce. Low global demand. US vulnerable.	Big investment required to close gap.	Strategic demand drives infrastructure build& incentives.	Only establish capability if government funds.	Breakthrough technology
High	Low	Low	Low	US needs; no one produces; invention required.	Basic research	Strategic demand drives research	Establish infrastructure to demonsrate technology & meet national demand	Conduct basic research
Low	High	High	High	US leads; strong exporter. Industry drives research based on global demand.	Applied research	Incentivize exports	Industry leads research & invests in production.	Breakthrough technology
Low	High	High	Low	US leads; strong exporter. Industry consortium leads future roadmapping.	Basic to applied research	Incentivize exports	Industry defines roadmaps, develops technologies and establishes infrastructure.	Conduct industry funded basic & applied research
Low	High	Low	High	US not global leader. Commoditized market.	Big investment required to close gap.	Unless US vulnerable, no action required.	Only invest if breakthrough enables global competitiveness.	Breakthrough technology
Low	High	Low	Low	New field. High export potential. No global Leader. New technology & infrastructure required.	Basic Research	Incentivize exports	Drives research & infrastructure investment. Partners with universities to conduct basic research.	Conduct industry funded basic research
Low	Low	XXX	XXX	No demand. Don't do it.				

4. The next step in the process is to create roadmaps in the broad technology areas identified in the national strategy. Two scenarios are considered for creating technology road maps:
   a. For mature industries, the preferable mechanism is to drive the roadmapping exercise through industry-government-university consortia.
   b. For nascent industries, since consortia will likely not exist, or will not have the scale needed to drive this exercise, the Federal Government will need to set up working organizations with broad participation.

   This also has to be an inclusive process utilizing collective intelligence as much as possible, with the aim of driving consensus. Here again, different mechanisms—such as expert panels, technology white-paper solicitations, surveys, competitive analyses white papers, data mining, workshops, etc.—have to be used. A process similar to the one outlined for the foresight process may be used, though face-to-face brainstorming workshops will probably play a larger role.
5. There are a number of recommendations in the area of program management:
   a. Programs need to be multiyear with stable funding because a certain mix of basic research goals has to be set, which requires time to yield results.
      Moreover, practical results and business impacts also take a few years to manifest themselves.
   b. It is critical that a co-funded model is used where both industry and government bring in the necessary components—the specific percentages will depend on the nature of the roadmap objectives—so that industry has the necessary participation and incentives to drive commercialization of technologies.
   c. For mature industries, consortia should create and manage programs. For nascent industries where the government plays a larger role in driving research and infrastructure and is therefore the primary stakeholder, programs will have to be managed by dedicated program managers from Federal agencies. Overall, many ongoing programs already exhibit best practices. For example, the DARPA practice of using short-tenure professional program managers may be a model worth emulating.
   d. As is largely done today, competitive-bidding processes need to be used for disbursing research funds for projects. Utilizing the gate-oriented approach of NIST ATP is recommended. In addition, clear metrics for proposal evaluation and award need to be set up. Broad criteria are recommended across all programs:
      i. Novelty of the approach
      ii. Impact of the approach in terms of enhancing manufacturing competitiveness by enhancing tradability and differentiation (e.g. in terms of reducing end-to-end supply-chain costs, reducing capital investments and risks, enhancing product features with significantly reduced costs).
      iii. Degree of addressing pre-competitive technologies, which should enable the industry and not a single player.
      iv. Degree of usage of educational and shared-research infrastructure.

      In addition, applied research programs should also include the following criteria:
      i. Business case in terms of time line of commercialization, potential market revenues, and ROI.

ii. Technical readiness of proposed approach (should be sufficiently mature).
iii. Sustainability of the approach.

Basic research programs can include the following additional criteria:

i. Potentially transformative nature of outcome (game changer, or opens up completely new areas of application).

e. Program policies also need to clearly lay out IP access rights for industry. Since industry co-funding is mandated, either exclusive-use or shared-use privileges need to be provided to all industry participants (depending on how the funds are actually allocated to individual projects).
f. Periodic reviews of the portfolio of projects need to be carried out by key stakeholders. Funding for programs need to be stable, but allocations within the portfolio need to be reviewed and adjusted. Portfolio allocations need to take care of the right mix of:
    i. Near-term benefits perhaps connected strongly to brown-field opportunities.
    ii. Longer term possibilities to create entirely new manufacturing opportunities—whether they are for products or manufacturing processes—as well as to help protect brown-field competitiveness.
g. Reviews need to ensure research has not been preempted by market forces and that it remains relevant to market needs. Rigorous, metric-based portfolio analysis needs to be incorporated by the program. Some suggested metrics are:
    i. Number and extent of "insertions" of research into products (or at least adoption into development programs in industries or as venture capital/private equity (VC/PE) funded investments).
    ii. Revenue and global market share generated by research (in terms of product revenues), which are more applicable for applied programs.
    iii. Number of patents.
    iv. Number of publications in important peer-reviewed journals.
    v. Number of students trained on the technology developed.
    vi. Number of SMEs benefited, or new startups spawned by, technologies.
    vii. Number of jobs created.

6. Programs need to leverage community colleges, universities, national labs, intermediate technology institutes (e.g., manufacturing innovation institutes) and independent research institutions (e.g., EWI) to carry out research and develop the talent pipeline for industry. Universities and national labs will typically be engaged by larger companies (and consortia) and for a larger component of basic research. The distinctions between universities and national labs should also be worked out to emphasize their complementary and partner roles and to avoid competitive concerns. In any case, it is critical that clear and sufficient goals of commercialization be set down in program objectives. In general, early-stage technologies need to leverage universities and national labs as shared infrastructure, whereas applied research has to revolve around the intermediate institutes and nonprofit research centers. In terms of the technology areas that have been explored as part of this study, some good areas that can be taken up by intermediate institutes/independent research institutes are sustainable manufacturing, digital manufacturing, information technologies, additive manufacturing, advanced forming and joining/bonding technologies, advanced manufacturing equipment, industrial robotics and some aspects of flexible

electronics. These fields already have a high percentage of applied research opportunities that need to be commercialized.

7. It is critical that SMEs and elements of the extended value chain are also involved and gain the required access to research infrastructure. Here, the intermediate institutes (MII) and EWI-like research institutes can play a vital role. Each institute will typically focus on a set of specific technologies. A mix of contract-based (i.e., for-fee project research for individual companies) and Federal funds needs to be deployed. The NISTMEP program can play a vital role in connecting SMEs to the requisite research resources in their areas of interest. State and regional involvement is also critical.
8. Industry further needs to co-fund internship and fellowship opportunities in universities and intermediate technology institutes to incentivize students to take up a manufacturing career path and provide them the right exposure to applied problems. Funding members or consortia need to be involved on advisory boards of universities/institutes to drive curricular changes to include new technologies that are developed. In general, early-stage technologies (low-technology readiness) will be present primarily in university curricula. Technologies that are intermediate in terms of their readiness level will typically be covered in both university and intermediate institutes as part of their applied-research programs. Very mature technologies will typically be the domain of community and technical training colleges as part of their curricula.
9. Finally, to ensure continuity through the whole technology life-cycle process and to be able to measure strategy through to execution, the National Program Office has to set up dedicated tracking teams for each broad technology area. The teams will need to link up the various activities and resources listed above and will essentially track progress and issues until the next cycle of the national strategy process. The teams also need to ensure sufficient cross-integration of the technology areas. All of this will facilitate a much-needed PDCA (i.e. Plan-Do-Check-Act) cycle. In fact, this "tracking" mechanism can be broadened to also function as a resource "web" that can speed up connections across all the different players in the value chain associated with the broad technology area, such as allowing SMEs to locate research resources, connecting early-stage technology developers with commercialization entities, etc.

Top Technology Areas

Based on the inputs and analysis carried out by this workstream, 10 technology areas have been selected as a starter list on which the stakeholders can focus:

- **Advanced sensing, measurement, and process control (including cyber-physical systems) also known as smart manufacturing, also known as advanced automation.** This set of cross-cutting technologies has applicability across almost all industry domains. These technologies are critical for enhancing tradability by way of end-to-end supply-chain efficiency (e.g., low-cost and pervasive sensors in plants and logistics systems, automatic control and coordination of systems of systems). In addition, megatrends of energy/resource efficiency and better safety/quality also depend highly on advances in sensing and automatic process control. Finally, emerging technologies such as nano-manufacturing or bio-manufacturing need their own specialized sensors and control models.

- **Advanced material design and synthesis.** This area covers design and synthesis of small molecules, nano-materials, formulated solutions, polymers, metals, fibers, coatings, composites, and integrated components (e.g., photovoltaic [PV] devices). This design and synthesis entails integration of computational modeling, state-of-the-art synthesis tools (e.g., high throughput), and advanced research analytics. Almost all the megatrends depend heavily on advanced materials, whether for energy efficiency, alternate energy devices, new materials to counter current resource shortages, next- generation consumer devices, or new paradigms in safety and security. Advanced materials will fuel emerging multi-billion dollar industries.
- **Visualization, informatics and digital manufacturing technologies.** This area includes integrated, enterprise-level smart-manufacturing methodologies (e.g., moving directly from computational/digital design to materials planning/purchasing to manufacture of customized formulated solutions for coatings to electronic materials). One aspect deals largely with manufacturing competitiveness through end-to-end supply-chain efficiency—reduced manufacturing cycle time, lower worker injury and illness rates, higher process yields, and higher energy efficiency, brought about by more networked information, control, and management across various entities in the value chain spanning across enterprises. The other aspect deals with the speed with which products are brought to market, which will be a key differentiator. It entails research focused on embedded sensing, measurement, and control systems for highly corrosive, high temperature processes affecting everything from PV to lightweight materials to polymer synthesis. It also entails control systems enabling manufacture of high-performance, highly controlled structures and devices. It also entails simulation and visualization technologies that can optimize the product and its manufacture in the virtual space (therefore bypassing time-consuming and expensive physical testing and experimentation) before actual physical production is started.
- **Sustainable manufacturing.** A key national need, this area covers high-performance catalysis, novel separations (including smart solvents), fluid mechanics, reactor design, etc. A major area of focus will be energy-efficient manufacturing, where high energy- consuming manufacturing processes need to be replaced with lower energy-consuming alternates. Areas such as remanufacturing (i.e., using recycled components to manufacture) also need to be researched. In addition to savings in energy consumption and higher profitability, many accompanying benefits can aid the competitiveness of industry. For example, an important mechanism to achieve energy efficiency would be to achieve a lossless process, which has quality benefits.
- **Nano-manufacturing (includes micro-feature manufacturing).** Nano-materials will most likely play a game-changing role in most future megatrends. Applications range from higher efficiencies in solar cells and batteries, to environmental control through nanotech-based filters, to nano-biosystem-based medical applications, to next- generation electronics and computing devices, to significantly enhanced material properties with nano-scale additives. Similarly, microstructures on devices

will play a key role in delivering new features or enhancing current functionality. The challenge will be to scale up volumes and reduce costs.

- **Flexible Electronics.** These technologies will be major differentiators in the next generation of consumer and computing devices. Some application industries such as photovoltaics and flexible displays are slated to be among the fastest growing ones over the next decade.
- **Bio-manufacturing.** Health care is going to be a key megatrend worldwide, with requirements for newer, more effective and less expensive molecules. Food security is also going to be a key concern of the future, where again bio-manufacturing will play a critical role. In addition, this technology has the inherent potential to enable energy efficiency in manufacturing. For instance, it offers room-temperature synthesis routes that can possibly replace current high-temperature processes. Innovations in the Bio– Nano-interface such as bio-inspired manufacturing using self-assembly have the potential to simplify and scale up many complex/expensive nano-manufacturing technologies and make them economically viable. According to MIT's Industrial Performance Center (IPC), bio-manufacturing "is a technologically advanced, innovative industry that requires highly skilled workers with commensurately high pay," and this industry can therefore play a vital future role in the economic value chain.
- **Additive manufacturing.** A possible megatrend for the future may be the production of highly customized/personalized products. Additive manufacturing ("3D printing") is the main technology that holds this promise. The technology also has several characteristics that enable other unique capabilities. For example, multiple materials can be processed, enabling smart components to be fabricated with embedded sensors and circuitry. Internal features can be produced that significantly enhance performance and therefore differentiate products, such as internal cooling channels optimized for thermal performance not possible with current manufacturing techniques. Also, materials can be processed efficiently with little waste, enhancing the sustainability of organizations that adopt additive-manufacturing technologies.
- **Advanced manufacturing equipment (including testing).** The national innovation ecosystem requires continuous feedback from challenges and new-requirements manufacturing. Securing a hold on the market for manufacturing equipment required in different industries (which is always highly capital intensive) ensures this feedback in a sustained fashion. In other words, even if specific plants are set up overseas, ensuring they have U.S. equipment ensures the necessary connectivity with ongoing manufacturing to close the feedback loop for continued innovation. In addition, these are advanced technology areas requiring significant research activity and funding, where maintaining differentiation is relatively easier for U.S. manufacturing.
- **Industrial robotics.** Automation and use of industrial robots in labor-intensive manufacturing operations such as assembly, product inspection, and testing can enable high endurance, speed, and precision. This has great potential to enhance productivity of the U.S. workforce and enable it to compete with low-cost economies, both for domestic and export markets.

- **Advanced forming (including near-net-shape manufacturing) and joining/ bonding technologies.** Most current mechanical manufacturing processes continue to largely depend on traditional technologies such as casting, forging, machining and welding. These technologies will most likely be the mainstay of future production processes for some time to come. But new needs for greater energy efficiency, resource efficiency, and greater performance require continued innovation and the search for disruptive technologies that will in turn help maintain U.S. competitiveness.

To summarize, a starter list of technology areas has been identified that can be used to kick-start the national strategic-planning process. These technologies address key national needs such as defense, energy independence and efficiency, food security, homeland security, and health care, and will have great bearing on ensuring U.S. manufacturing competitiveness in terms of both differentiation and tradability of products. Because of the interplay beetween these technologies, they need to be developed in tandem to ensure greatest impact.

Current support from an illustrative set of Federal programs for these technology areas is shown in the table below. Although a number of these programs are well aligned with industrial needs, it likely that gaps remain in the Federal portfolio. Steps should to be taken to carry out a complete analysis of the portfolio and to ensure additional investment in the areas where alignment is lower than required.

Advanced Technology	Agency	FY11 and FY12 Programs
Advanced sensing and measurement technologies	NSF	Cyber physical systems
Nano-manufacturing	NSF	Nano-manufacturing
Information technology		
Sustainable manufacturing	DOE EERE ITP	Industrial energy efficiency ($120 MM/ 3 years)
Nano-scale materials	DOD	Alternate energy
Continuous process control		
Flexible electronics	DOE, NIST	Flexible electronics for batteries and solar cells, NITS (TIP)
Process control		
Visualization		
Adaptive control		
Coatings		
Bio-manufacturing	DOD/ DOE/NIST	Bio-manufacturing; low-carbon biosynthesis of industrial chemicals (% $500 MM), TIP program (NIST)
Biofuels	DARPA	Synthetic biology ($35 MM)

Advanced Technology	**Agency**	**FY11 and FY12 Programs**
Lightweight materials	DOD/ DOE	Small, high-powered batteries; cost-effective, ultralight, ultra-durable materials for autos (% $500 MM)
Material genome	DOD/DOE NIST/NSF	Materials Genome Iniative ($100MM)
Optoelectronics		
Precision machining	DOD	Metal fabrication
Recycle waste-management technologies		
Simulation/test infrastructure	NSF	Cyber physical systems—smart buildings and bridges
Additive manufacturing	NSF	Advanced manufacturing (manufacturing machines and equipment)
Advanced ceramics		
Composite assembly	DOD	Advanced composites
High-temperature processing		
Industrial robots		
Nano-technology medical diagnostics devices and therapeutics		
Ceramics	DOD	Transparent armor, stealth technology ($24MM)
Conductive inkjet technology		
High-speed mixing		
Mobile robots	SBIR/NSF	National Robotics Intiative ($70MM)
Reaction engineering		
Separation technologies		
Metal-jet technology		
Ultra-efficient waste heat recovery		
Advanced forming (including near-net-shape manufacturing)	NSF	Advanced manufacturing (manufacturing machines and equipment)
Advanced joining/bonding technologies	NSF	Advanced manufacturing (manufacturing machines and equipment)

SUPPORTING MATERIALS

Sustaining Mechanism for Technology Development

Best-practice public-private partnership programs: Survey findings

	GUIde	AFRAPT	NRI	FCRP	SEMATECH	MAI	NREC	ATP	DARPA	ARPA-E	DOE-EERE	TAR DEC/NAC	CAPD
Objective	to improve bladed disk forced response and high cycle fatigue prediction in aircraft engines	Talent pipeline - advanced engineering degrees in areas such as aerodyne., combustion, and the structural dynamic sciences	Alternative computing devices to CMOS FET	enable ultimate CMOS technology scaling and enable highly complex designs	enable ultimate CMOS technology scaling and enable highly complex designs	Develop low cost metals for US airforce	To develop and mature robotic techs. from concept to commercialization	early stage, high-risk technology dev. that would otherwise go unfunded	pursue and exploit fund. science and innovation for National Defense	To focus on creative "out-of- the-box" transform.l energy research that industry by itself cannot or will not support due to its high risk but where success would provide dramatic benefits for the nation	invests in clean energy technologies that strengthen the economy, protect the enviro., and reduce depend. on foreign oil	deliver the most advanced technology solutions to improve the Nation's ground vehicle fleet	Under. and aid complex design and operation issues faced by industry Develop and advance modeling and solution methods for process systems eng.
University participant?	Y	Y	Y	Y	Y	Y	Y	Y	Y	Y	Y	Y	Y
Consortium	GUIde	N	SRC	SRC	SEMATECH	MAI	N	N	N	N	N	N	CAPD
Best practice criteria													
Funding Period	15 yrs	Multi-year	Since 2005	Since 1999	Multi-year	Annual	Multi-year	Multi-year	Multi-year	Multi-year	Multi-year	Multi-year	Multi-year
Clear mission	Y	Y	Y	Y	Y	Y	Y	Y	Y	Y	Y	Y	Y

Best-practice public-private partnership programs: Survey findings (Continued)

		GUIde	AFRAPT	NRI	FCRP	SEMATECH	MAI	NREC	ATP	DARPA	ARPA-E	DOE-EERE	TAR DEC/N AC	CAPD
Indus try role	Program Mgt	N	N	Y	Y	Y	Y	N	N	N	N	N	N	N
	Research direction	Y	Y	Y	Y	Y	Y	Y	Y	N	Y	Y	N	Y
	Project steering	Y	N	Y	Y	Y	Y	N	N	N	N	N	N	Y
	Practical experienc e		Y	N	N	Y	Y	N	N	N	N	N	N	Y
	Co-funding	cost shared exper. hardware and on projects	Summer internships at engineer salaries	Cost shared	Cost shared	Cost shared	Cost shared	N	N	N	N	Cost shared	N	Cost shared
Research type		Applied rather than basic research	Applied research	Basic	Applied	Applied + basic	Applied	Applied	High risk - applied + basic	High risk - breakthru	High risk - breakthru	demo + basic	Applied + basic	Applied
IP provisionsfor industry		?	NA	Pre-comp.	Pre-comp.	Pre-comp.	Yes, with govt getting nonexcl. license	Licensed from NREC	Yes-compani es retain them	Yes(but with FAR clauses)	Yes, with govt getting nonexcl. license	Yes, with govt getting nonexcl. license	inventing party retains IP	?
Other success criteria			Paid students 30% higher than average				Business case oriented. Independe nt Technical oversight committee		Already had some proof of concept in place, gate based selection, biz case oriented	Strong program managers		expert panel to provide steering inputs,		

Comparative Study of Technology Development Mechanisms in other Countries

Germany

Germany continues to maintain a visible national strategy on manufacturing overseen by the government. The German Federal Ministry of Education and Research (BMBF) uses a foresight process:

- The BMBF develops a High-Tech Strategy (HTS) that lays out the key areas of research.
- Based on the broad fields derived from the HTS, a number of qualitative and quantitative methods are used (e.g., data mining, workshops, expert interviews, bibliometrics) to determine new areas of research.
- An international panel consolidates these ideas into a set of topic candidates.
- The BMBF then generates a set of white papers on these topics.
- These white papers and topic candidates are eventually distilled into a research agenda using a series of expert workshops and conferences, online surveys, and a comparative study of other innovation systems.
- Other more specific studies (also typically highly consultative and inclusive) are also initiated from time to time by the BMBF to augment the research agenda.
- The national ecosystem consists of the various universities and applied research institutes (e.g., the Fraunhofer institutes), as well as private companies and institutions.

German funding for research is mainly administered by the BMBF through the German Research Foundation (DFG). Most German government manufacturing research funds flow through the technical universities. But note that the Technical Universities carry out a high degree of applied research for industry clients. In addition, the Fraunhofer Institutes are solely applied-research institutes and mainly get their funding through contract work from industrial clients as well as state and federal agencies.

One additional key point is the importance of the SME segment (Mittelstand) and its inclusion in the national strategy on manufacturing. The ecosystem and research/funding mechanisms ensure that SMEs have access to applied research resources.

Japan

The Ministry of Economy, Trade and Industry (METI) generates the National Strategic Technology Roadmap (STR) every few years (with annual updates). A Delphi forecasting process is used where multiple rounds of input are carried out with a wide net of experts from government, industry, and universities, and the areas of research are gradually distilled to the final set. For the 2010 STR, over 800 experts were involved in the process.

Korea

South Korea also follows a Japan-like model where a National Technology Roadmap (NTRM) is created once every few years. This process is led by the Ministry of Science and Technology (MOST). The first instance of this national foresight process was carried out in 1994.

This was a much broader exercise where blank-slate inputs from more than 30,000 experts from industry, government, and academia were invited on research topics of importance. A main expert coordinating committee as well as 12 expert subcommittees were set up to coordinate the process and distill ideas from the huge set of responses and to create the main questionnaire for the Delphi forecast. The Delphi was carried out over 2 rounds with more than 4000 experts being involved.

Top Technology Areas

First Survey with Industry and Universities

The chart below describes the percentage of industry respondents who selected each industry as critical for maintaining U.S. manufacturing competitiveness.

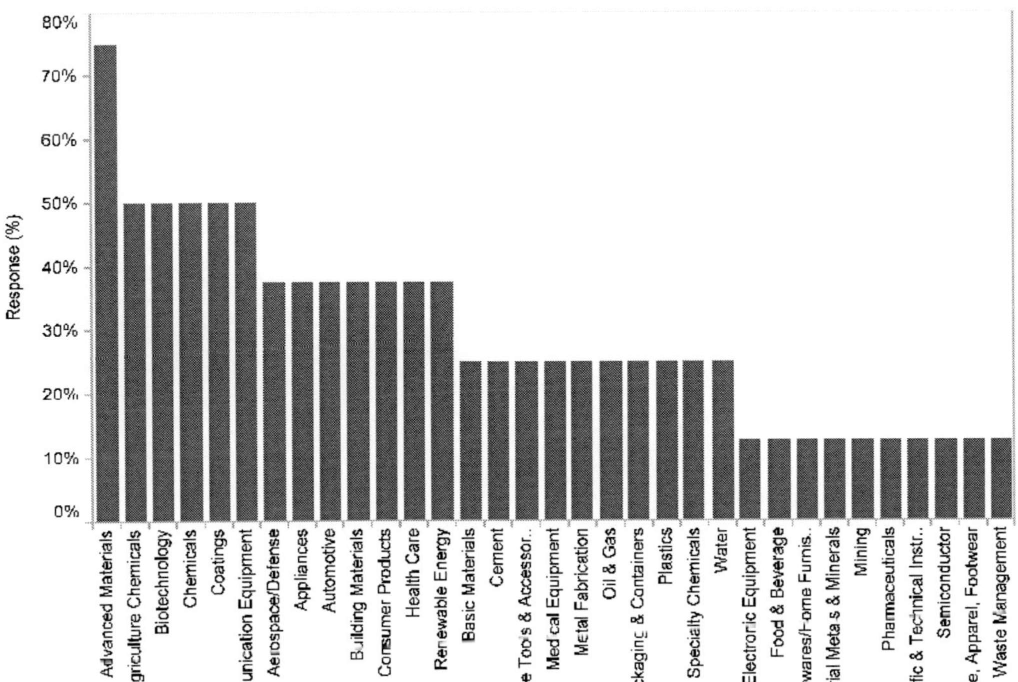

Industry Survey Results: Industries Targeted to Retain and Grow U.S. Manufacturing.

The chart below depicts the percentage of industry respondents who selected each advanced manufacturing technology as being critical for U.S. manufacturing competitiveness.

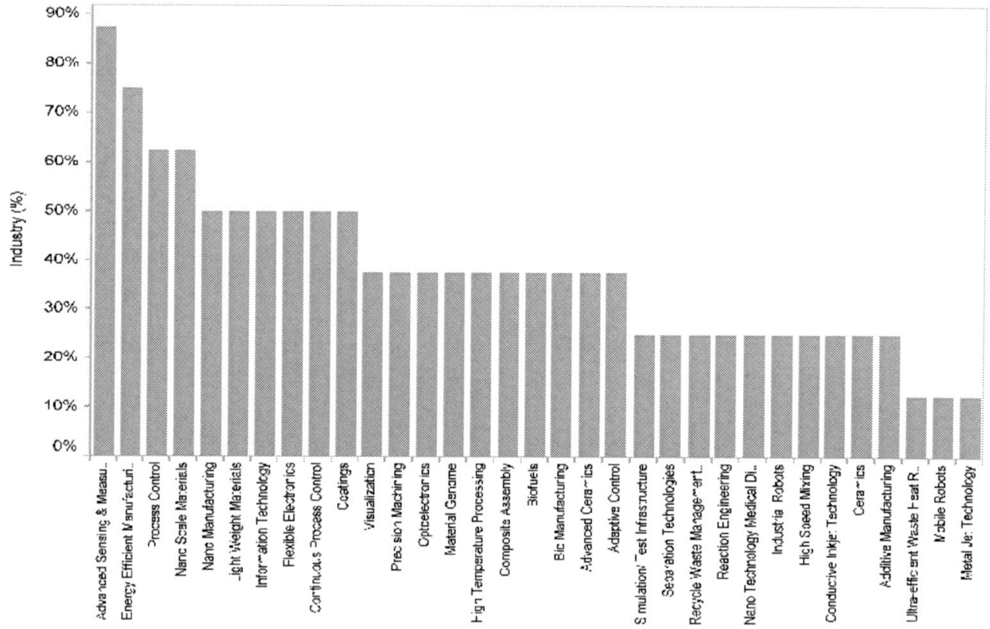

Industry Survey Results: Advanced Technologies Required to Retain and Grow U.S. Manufacturing.

The chart below depicts the percentage of university respondents who selected each advanced manufacturing technology as being critical for U.S. manufacturing competitiveness.

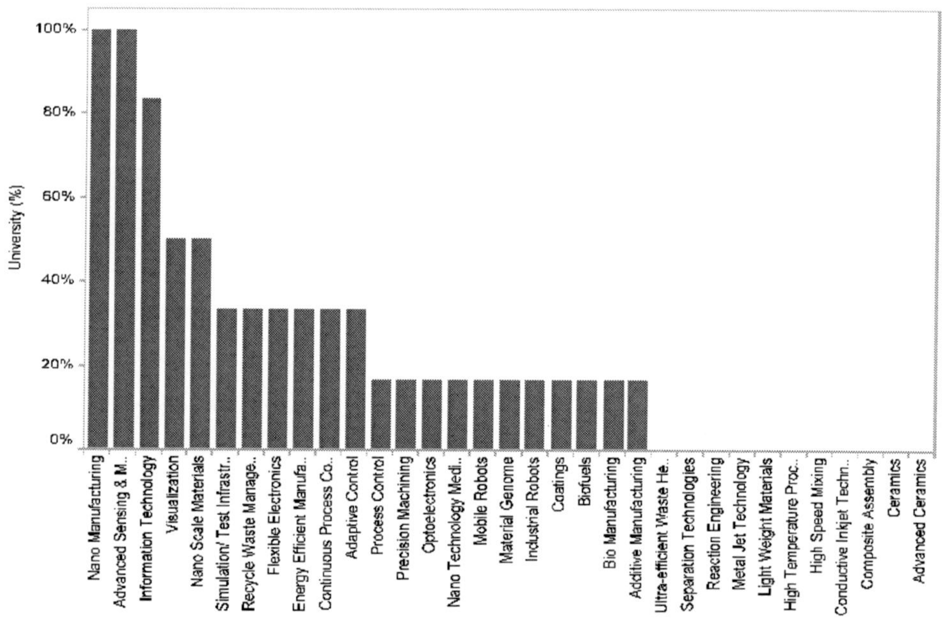

University Survey Results: Advanced Technologies Required to Grow U.S. Manufacturing.

The chart below combines the selections of industry and university respondents for advanced manufacturing technologies that are deemed critical for U.S. manufacturing competitiveness.

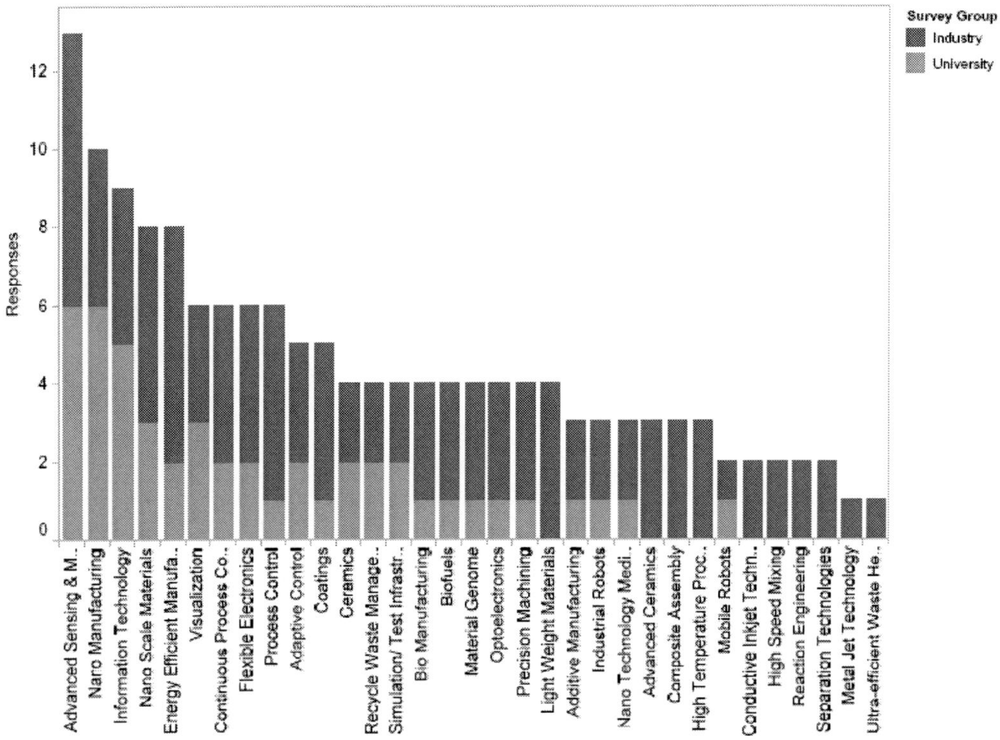

Combined Survey Results: Top Technologies Deemed Most Critical for U.S. Advanced Manufacturing.

Factors for Manufacturing Outside or Inside the United States

The first chart in the pair below shows the factors driving decisions to manufacture outside the United States, as identified by industry participants. The factors are ranked according to the frequency of selection by the respondents. The second chart of the pair shows the factors driving the decision to bring back manufacturing to the United States

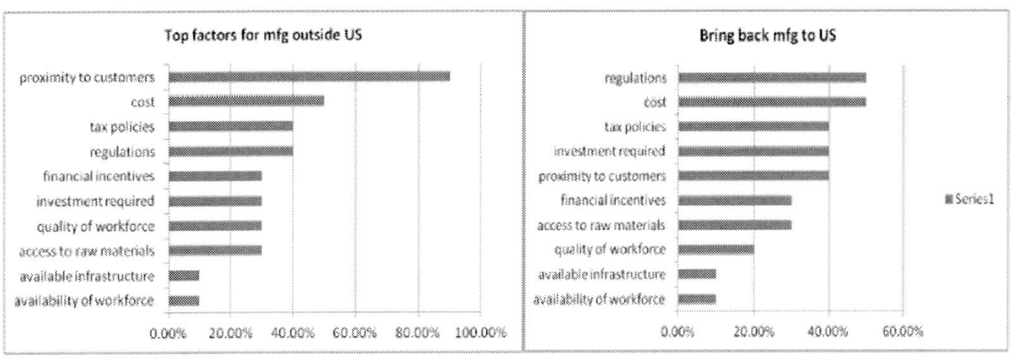

MAPI Council Survey: Excerpts

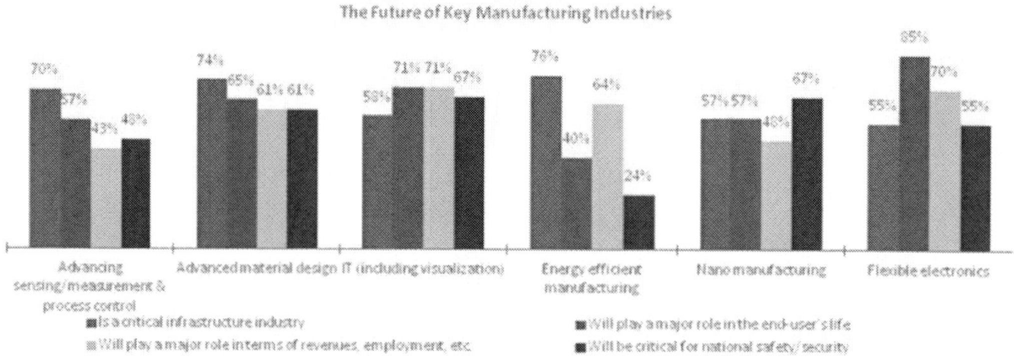

- Respondents rated advanced material design as critically important across the board.
- Advanced controls and energy-efficient manufacturing were viewed as critical infrastructure industries.
- Information technology (IT) and flexible electronics are both seen as vitally important to end users in terms of revenue and job generation.
- Over two-thirds of those who rated nano-manufacturing expect it to be critical for national safety or security in the future.

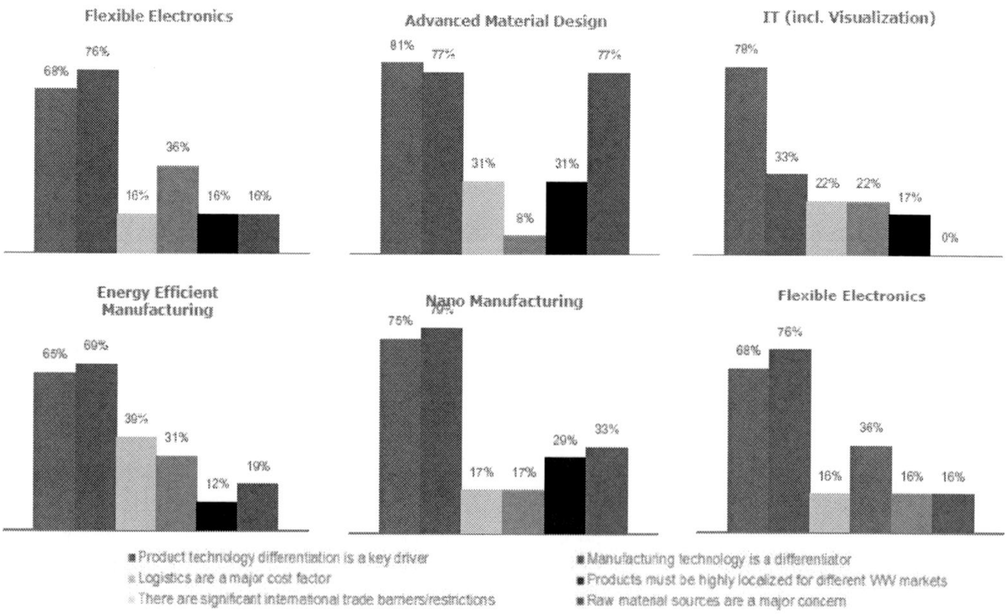

- A very high proportion of respondents identified product technology differentiation as a key driver for controls, material design, and IT.
- With the exception of IT, differentiation of product technologies is expected to be a key driver of success for advanced manufacturing.

- In general, adopting advanced manufacturing technologies is not expected to be seriously affected by logistics costs.
- Other than in energy-efficient manufacturing, the need for localized design is also not expected to play a major role in future development, perhaps because of increasingly globalized standards and consumer expectations.
- At this point, trade barriers are not expected to interfere with dissemination of advanced manufacturing technologies.
- Access to raw materials could present a challenge in the case of advanced material design.

Member Comments: "Our human resources need to be significantly more competitive. Knowledgeable and productive."

"Stability of the rules (tax, regulatory) more important than the rules and taxes themselves. These set a framework for the investment case but aren't drivers of the cases - setting a stable and competitive framework for companies to do business in is the most important failing of recent times."

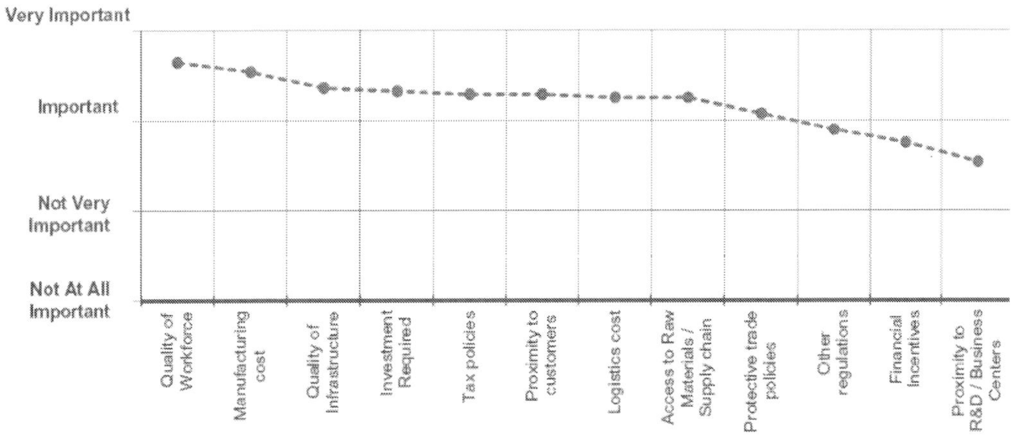

Importance of Key Factors for Deciding to Manufacture in the U.S.

Member Comments: "Determining your manufacturing footprint is a delicate balance between optimizing your manufacturing costs and your best tax rates. Everyone would prefer to manufacture in the US, but an inflated real wage rate and the 2nd highest corporate taxes in the world make it a bad financial decision."

"Providing most competitive landed costs at point of use for final assembly is key. If we can offset lower wages offshore with higher efficiencies in our US factories, the only reason for sourcing outside the US would be based on prohibitive regional transportation and logistics costs or market entry barriers or duties brought about by lack of trade agreements."

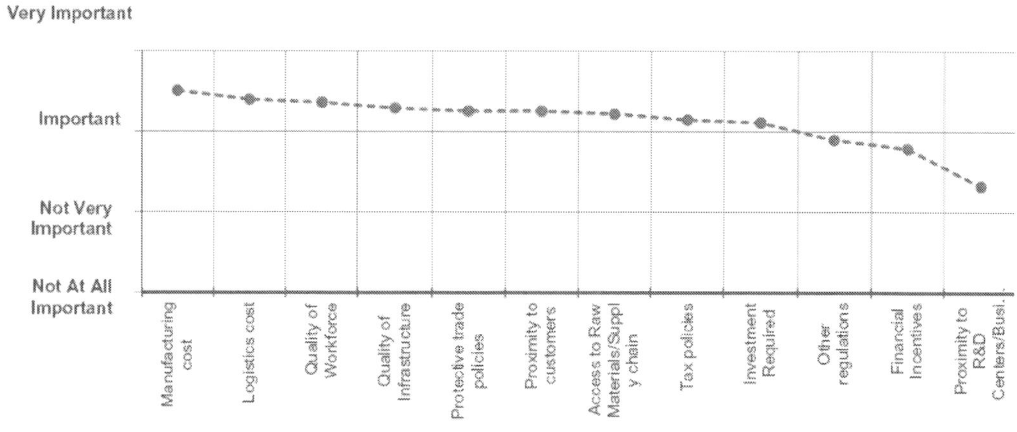

Importance of Key Factors for Deciding to Manufacture Outside the U.S.

References

[1] Eoin O'Sullivan. (2011). A review of international approaches to manufacturing research, *University of Cambridge Institute for Manufacturing*, March.

[2] PCAST Report to the President on Ensuring American Leadership in Advanced Manufacturing, June 2011.

[3] 2012 Global R&D Funding Forecast, Congressional R&D Caucus Presentation, February 2, 2012, Batelle/R&D Magazine

[4] Proposal for a COUNCIL DECISION establishing the Specific Programme Implementing Horizon 2020 - The Framework Programme for Research and Innovation (2014- 2020){SEC(2011) 1427-Volume 1}{SEC(2011) 1428-Volume 1}, 30.11.2011

[5] Taeyoung Shin, Technology Forecasting and S&T Planning: Korean Experience, Prepared for the Brazil International Seminar on Foresight Studies, September 2000.

In: Advanced Manufacturing in America
Editors: Brandon Cullerne and Michael Laston
ISBN: 978-1-62257-932-7

Chapter 4

CAPTURING A DOMESTIC COMPETITIVE ADVANTAGE IN ADVANCED MANUFACTURING. *ANNEX 2: SHARED INFRASTRUCTURE AND FACILITIES WORKSTREAM REPORT*[*]

President's Council of Advisors on Science and Technology

ABSTRACT

In June 2011, the President established the Advanced Manufacturing Partnership (AMP), which is led by a Steering Committee that operates within the framework of the President's Council of Advisors on Science and Technology. In July 2012, the AMP Steering Committee delivered its report to PCAST, entitled *Capturing Domestic Competitive Advantage in Advanced Manufacturing*. PCAST adopted this report and submitted it to the President. The Steering Committee's report draws on preliminary reports prepared by several "workstreams." These workstream reports have been made available as on-line annexes to the Steering Committee report.

EXECUTIVE SUMMARY

Upon extensive benchmarking and analysis of various shared infrastructures and facilities in the United States and abroad, the Advanced Manufacturing Partnership (AMP) Steering Committee's workstream on Shared Infrastructure and Facilities recommends two actions to improve the competitiveness of U.S. manufacturing: (1) establish a network of Manufacturing Innovation Institutes (MIIs) to bridge the gap between basic research performed at

[*] This is an edited, reformatted and augmented version of President's Council of Advisors on Science and Technology, dated July 2012.

universities and national laboratories and the work done at U.S. production enterprises; and (2) establish a National Advanced Manufacturing Portal to provide searchable catalog of data, services, and facilities that are made available through publicly funded cooperative research centers and laboratories located within a large number of U.S. universities and national laboratories.

The MIIs, a network of public/private partnerships, would serve as regional centers for (1) promoting collaboration among industry, academia, and government on applied research and development in emerging technology areas, (2) facilitating the quick adoption of new manufacturing technologies, tools, and methodologies to make U.S. manufacturing more competitive, and (3) developing technical workforce with training and experience required by industry.

The Portal searchable on-line catalog of publicly funded research centers, services, and facilities would enhance access to these centers and facilities by small- and medium-sized manufacturers.

CHARGE TO THE WORKSTREAM

The objective of the Shared Infrastructure and Facilities Workstream is to assess opportunities to de-risk, scale up, and lower the cost of accelerating technology from research to production through unique capabilities and facilities that serve all U.S.-based manufacturers, in particular small- and medium-sized manufacturers.

PROCESS FOLLOWED

The AMP Steering Committee Shared Infrastructure and Facilities Workstream conducted an extensive benchmarking of the various shared infrastructure and facilities in the United States and abroad. We first identified a set of key attributes for the benchmarking exercise, including mission/charge, partnership model/governance, membership composition (number of large-, small-. and medium-sized companies), sectoral reach, geographical reach, funding mechanisms, intellectual property (IP) agreement among parties, and metrics of success. We then identified a set of centers, institutes, and facilities from the United States and abroad for our benchmarking analysis. These entities included the National Institute of Standards and Technology (NIST) Center for Neutron Research, NIST Manufacturing Extension Partnership, National Science Foundation (NSF) Engineering Research Centers, Department of Energy (DOE) Innovation Hubs, California Innovation Hubs, National Nanotechnology Infrastructure Network, Semiconductor Research Corporation, Industrial Technology Research Institute of Taiwan, Advanced Manufacturing Centers in the U.K., and the Fraunhofer Institutes in Germany, among others. Key findings from the benchmarking and the desired attributes of an innovation infrastructure for advanced manufacturing are summarized in the section that follows.

KEY FINDINGS

The United States has been a global leader in research and discovery, enabled by our first-class research universities and national laboratories. However, many of our research discoveries have not been quickly translated into products or applications in manufacturing in the United States. Many technologies fail to move to commercialization because the private sector, and particularly small- to medium-sized companies, is not able to make sufficient investment in early technologies and the cost of prototypes and scale-up are high. In fact, the stage from research to production is a perilous period in business development that is often called "the valley of death." This problem is attributable, in part, to the significant differences in the way activities in research and manufacturing are conducted. Basic research and new discoveries tend to happen in a largely disorganized endeavor, with the end goal most often being to publish the results in scientific journals. By comparison, manufacturing activities are competitive and must be focused and systematic. Channeling the results of research creativity into manufacturing requires systematic translation, supported by an "intelligent blend of public and private sector investment, targeting the most promising technologies" [1] and facilitated by shared infrastructure and facilities. Several countries have done well in this regard. In fact, the PCAST Report to the President on Ensuring American Leadership in Advanced Manufacturing provides various examples of products that are "invented here, but produced elsewhere" [2]. These examples include e-readers, flat panel televisions, semiconductor production equipment, and lithium-ion batteries. Many of these high-technology products are produced in China, Korea, and Taiwan where the governments continue to provide critical support for early technology adoption, manufacturing, and commercialization.

Our benchmarking exercise reinforced many of the observations that have been published in the various reports and reveals a significant gap in the U.S. innovation infrastructure:

- U.S. universities receive a great deal of Federal funding for basic and applied research. Though there has been an increasing emphasis by universities on technology transfer and commercialization, only a small number of discoveries and findings are translated into new products or useful methods, processes, or software to enhance economic competitiveness. Most federally funded research results in publications in scientific journals.
- The business sector performs the largest portion of U.S. research and development (R&D) work, using internal resources [3, 4]. Several large corporations, including Dow, Ford, GE, and GM, have realized the importance of long-term partnerships with academia in research and education and have developed such partnerships. But total industrial support for universities remains limited and rarely do small and medium-sized enterprises (SMEs) attempt to fund research grants and contracts with faculty members in research universities.
- The funding mechanisms for transferring the research findings and discoveries into tangible new products and manufacturing applications (e.g., NSF Small Business Technology Transfer [STTR] and Grant Opportunities for Academic Liaison with Industry [GOALI]) have been limited in their effectiveness to enhance manufacturing competitiveness due to the limited scale and duration of support.

- The United States lacks strong, branded intermediary institutes focused on applied R&D activities that bridge the gap between research and manufacturing.
- Large companies have developed staffs to produce the modeling and simulation software they need for competitive advantage, but the vast majority of SMEs are not capable of developing, acquiring, or using such software [5].
- The European Commission has demonstrated (through a Fraunhofer COVES Center) that SMEs can greatly improve their performance when they are provided appropriate assistance and use of modeling and simulation (M&S) capabilities [6].
- The new National Digital Engineering and Manufacturing Consortium (NDEMC) at Purdue University has provided M&S software to a few U.S. SMEs, which has allowed them to successfully compete against foreign manufacturers [7]. But only a few universities and software companies are forming collaborative, multi-disciplinary teams that can produce the M&S software tools appropriately configured for SMEs to use.
- SMEs have a difficult time (1) finding what resources are available to them and (2) accessing those resources once they find them.
- The lack of real-world applied problem-solving experience by faculty and students in research universities has led to companies hiring college graduates that still require extensive training to function well in a company.

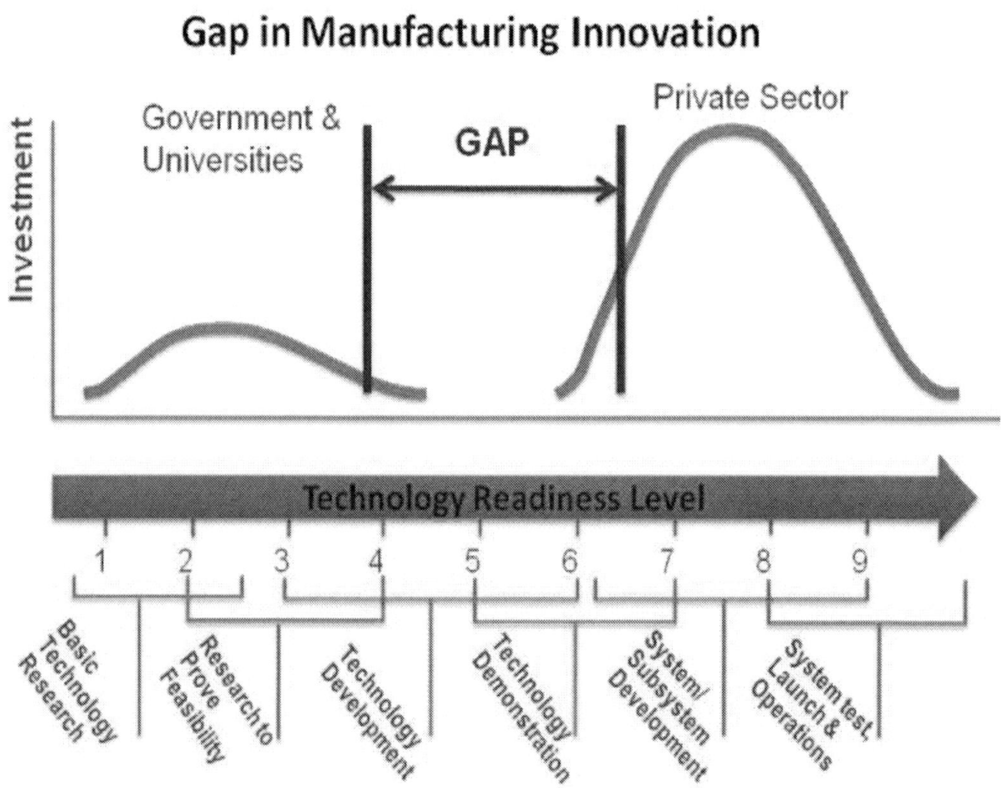

Our benchmarking also identified several key desirable attributes of a shared national infrastructure for supporting the translational activities for bridging fundamental research and manufacturing:

- Long-term partnership between industry and universities, enabled by government
- A sustained focus on technology innovation with a strong brand identity and reputation
- Ability to identify critical emerging technologies with transformational impact and capacity in translating these technologies into products and businesses for the market
- Ability to form effective teams of industrial and academic experts from multiple disciplines to solve difficult problems
- Dual appointments of faculty/students in both research universities and application-oriented institutions with access to fundamental research as well as opportunities for applied problem-solving to develop leaders and a workforce equipped to deal with the new technologies and production systems
- Ability to engage and assist small- to medium-sized companies that need new technologies

RECOMMENDATIONS

Manufacturing Innovation Institutes

Advanced manufacturing is defined by PCAST [2] as a "family of activities that (a) depend on the use and coordination of information, automation, computation, software, sensing, and networking, and/or (b) make use of cutting edge materials and emerging capabilities enabled by the physical and biological sciences, for example nanotechnology, chemistry, and biology. This involves both new ways to manufacture existing products, and especially the manufacture of new products emerging from new advanced technologies." A national network of Manufacturing Innovation Institutes (MIIs) should be established to support advanced manufacturing. The initial areas of MII support would be in manufacturing technology areas recommended by the AMP Steering Committee Technology Development Workstream. Future areas of support could be expanded to include areas of emerging technologies that have the greatest potential for translation into products and businesses. These areas are to be identified through a regular advanced manufacturing technology roadmap process, as described by the Technology Development Workstream. (See Annex 1 for a recommended approach to develop this permanent model and roadmap process). An open, competitive process with peer reviews ought to be used to establish the MIIs.

The goals of the MIIs are to promote collaboration among industry, academia, and government on applied R&D, to address emerging technology areas where market failures are causing U.S. innovations to be scaled up and manufactured elsewhere, to facilitate the quick adoption of new manufacturing technologies, tools, and methodologies that will make U.S. manufacturing more competitive, and to develop technical workforce with training and experience required by industry.

We recommend that each institute:

- Focus on an area of U.S. national economic strength or a promising emerging technology.
- Be supported by a mixed funding model with government funding being guaranteed for a minimum of 5 years with the potential of renewal for a total of 10 years, to allow for long-term project development and the ramp-up of private sector support.
- Be hosted by an industrial consortium, a university, or a national laboratory. A new or existing partnership would be eligible to apply for Federal Government funding with demonstrated commitments from industry, a state government, and a research institution. A partnership must have among its members a minimum of two large companies and shall have participation of related small- and medium-sized companies, and at least one major research university along with other regional universities and community colleges.
- Be governed by a Board of Directors composed of representatives from business, academic, and government organizations supporting the MII.
- Operate independently with contractual flexibility, but all MIIs will be members of the national network and will follow a similar governance model defined by a national governing board.
- Be staffed with full-time applied researchers who are experienced in bringing research into production, innovation enablers who support the process of technology identification and commercialization, short-term contract researchers who have specialized expertise, industrial scientists and engineers in residence, and part-time faculty, post-doctoral researchers, and student interns.
- Establish distributed manufacturing support centers throughout the region to support small- and medium-sized manufacturers that may adopt new technologies.
- Provide assistance to community colleges that seek to develop advanced manufacturing programs.
- Provide grants to other universities and businesses that are developing complementary and enabling technologies.

These MIIs would provide a shared infrastructure for technology development and serve as a "collaboratory" between research universities and businesses by providing existing and start-up businesses with greater access to research, students, internships, workforce training and development, technology transfer, and commercialization. They would also provide a variety of business services such as design, digital manufacturing, prototype and test services, and staff training.

A national manufacturing innovation infrastructure of this type would strengthen U.S. economic competitiveness in several fundamental ways:

- New technologies would not only continue to be invented in the United States, but many of them would be translated quickly into new products produced here, because the MIIs would reduce the risk of development and production through public/private partnership and shared facilities.

- Existing manufacturers would become more competitive as new manufacturing technologies, tools, and methods are transferred more effectively to production applications.
- Training of college graduates and re-education of industry workforce would be more relevant and responsive to the needs of manufacturers.
- New jobs would be created around specific technology clusters that are created and commercialized.

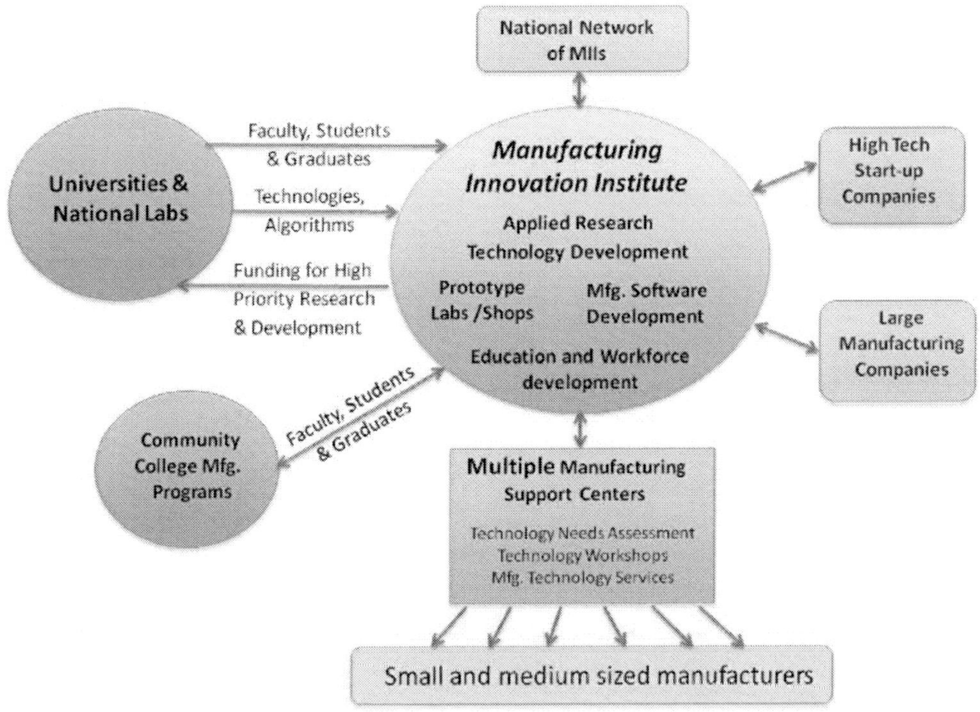

Funding for MIIs

Long-term government support for the MIIs would be necessary as it serves as a catalyst for long-term public/private partnerships. Private sector support, consisting of membership, contract projects, and revenues from commercialization, should be about a third of the total MII annual budget at steady state. This amount is to be matched by Federal Government funds. The remainder of the funding will be from state and university sources and other competitive grants. The Federal Government may provide a more significant portion of support during the launch and ramp-up stage of the MIIs.

Partnership: The membership fee structure for an MII shall be determined by its governing board. Membership fee and government support shall be used to support research projects of common interest. IP resulting from such projects shall be jointly owned by the members of the partnership. All members would have the option to acquire a nonexclusive, royalty-free license, in a field of use chosen by the member without the right to sublicense, to the patentable results of MII funded projects.

Contract research: An MII may engage in fee-based contract research and development with member or non-member companies. Rights to the IP from such contract research shall belong to the paying companies. Contract research and development may leverage shared infrastructure enabled by government support.

Performance Review for MIIs

The MIIs shall follow a regular schedule of independent reviews based on a set of criteria established by the MII network, which may include inventions and other IP, technology licenses, startups, companies (in particular, SMEs) supported, and company satisfaction. In addition to the annual review, a major review shall be carried out every 3 years. After a 3-year major review, an MII may be recommended for funding for another 5 years upon successful review (a second 3-year review will be carried out during year 6 or be recommended for closing within 2 years upon a negative review. An MII may re-compete for government support every 10 years if the MII adopts new technology focus areas.

Manufacturing Innovation Institutes Compared to Existing Research Entities

The MIIs represent a coordinated national network for supporting translational activities to bridge the gap between research and manufacturing. Translational research will be the hallmark of the new innovation infrastructure, with the primary measure of success being the number of new high-value products being manufactured by U.S. companies in the United States and new advanced manufacturing processes and technologies being adopted. The MIIs may be compared to the following existing research entities:

- *NSF Engineering Research Centers (ERCs)*: The ERCs are charged with conducting science-based research in creating the next generation of engineering systems and educating students at all levels about the science and technology of such systems. Activity within an ERC lies at the interface between the discovery-driven culture of science and the innovation-driven culture of engineering. Industry collaboration is required, but membership fee structure is not mandated by NSF, and industry serves in an advisory role (Industry Advisory Board).
- *NSF Industry and University Cooperative Research Centers (I/UCRCs)*: NSF began the I/UCRC program in 1973 and approximately 50 I/UCRCs have been supported by NSF. These are membership-based centers with NSF providing a small amount of support, ranging between $35 thousand and $70 thousand annually. To qualify for NSF support and I/UCRC designation, the membership must consist of six companies, with a total annual membership fee of $300 thousand. Industrial members serve on the Advisory Board. The types of research projects carried out in I/UCRCs vary greatly from center to center.
- *DOE Innovation Hubs*: The concept of the Innovation Hubs was based on the Discovery and Innovation Institutes developed by James Duderstadt through the Brookings Institute. While the hub model is of value to the MIIs, the key difference

lies in that the hubs are focused on advancing promising areas of energy science and engineering from the earliest stages of research to the point of commercialization and are funded entirely by the Federal Government. The Hubs also do not have a mandate to improve the technological capabilities of SMEs.

Example Technology Areas to Establish Manufacturing Innovation Institutes

The technology areas that we recommend be supported by the MIIs have been identified by the AMP Steering Committee Technology Development Workstream, and new ones would be identified through the regular Technology Roadmap exercise, as recommended by the Technology Development Workstream. Several technology areas that have received enthusiastic support from our regional meetings and industry surveys are:

- *Lightweight Structures:* Composites, titanium, and other materials have wide applications in aerospace, automotive, and the defense industries. Cost-effective manufacturing of such materials into lightweight structures can lead to enhanced product performance and reduced energy consumption. An MII focused on developing innovative, cost-effective processes for these materials and the joining and assembly of them into structures would have a significant impact on performance and energy efficiency of commercial and defense products.
- *Manufacturing Scale-up for Flexible Electronics:* Electronic circuits mounted and assembled on flexible substrates allow them to be reshaped and bent during use. Flexible electronics have many applications, including 360 degree cameras, sensors, health monitors, electrical connections between subassemblies and the like. An MII focused on the development of scalable production technologies for flexible electronics would lead to broad applications of flexible electronics, leading to new products and businesses.
- *Digital Manufacturing:* Advanced simulation and modeling technologies enable manufacturers to predict product and manufacturing system performance with such great fidelity that they no longer have to build and test costly physical mock-ups of proposed new products, manufacturing processes, and facilities, which, in the past, resulted in added costs and long delays in bringing new products to markets. Unfortunately, small- and medium-sized companies, who comprise the supply chains to large manufacturers, do not possess such modeling and simulation capabilities. At the same time, new modeling and simulation algorithms are constantly being developed in universities, but they are not directly usable in manufacturing. A Digital Manufacturing Innovation Institute could be set up as one of the first MIIs, with its missions being: (1) selecting, evaluating, and certifying existing manufacturing-related engineering modeling and simulation software; (2) providing grants to commercial software developers to provide "pay-as-you-use," cloud-based high-performance computing (HPC) software needed by SMEs that want to use simulations of existing and future manufacturing processes, materials, and operations; (3) providing funding to other universities and laboratories for the computational modeling of current research results related to new materials, manufacturing processes, and operations; and (4) supporting SMEs in digital

manufacturing. Such an MII would help improve U.S. manufacturing competitiveness by shortening the time to market and improving quality and productivity of manufacturing.

National Advanced Manufacturing Portal

The goal of the National Advanced Manufacturing Portal is to provide a searchable catalog of data, services and facilities that are made available through publicly funded cooperative research centers and laboratories located within a large number of U.S. universities and national laboratories, in order to enhance access to such resources by small and medium-sized manufacturing firms in the U.S.

The proposal for a National Advanced Manufacturing Portal was developed in the spirit of addressing government coordination and data access issues recently highlighted by President Obama. Firms as well as industry and technology experts reported that conventional web searches (like Google) did not produce useful results. This problem is in part due to the vast variation and complexity of the research and innovation conducted throughout the U.S. network of cooperative research centers. Simply put, finding a practical answer to the simple questions that pervade advanced manufacturing such as, "Where can I find out if there is a better adhesive that works just as well as the one I use now?" or "Is there an alternative to this film coating?" is nearly impossible for small firms with limited time and limited R&D staff.

The National Advanced Manufacturing Portal would address this problem by creating a single web portal where SMEs (as well as others) can search for the cooperative research centers that best meet their needs. With this information, firms can make both short and long-term R&D plans. The Portal would make progress towards the goal of "pushing innovation down the supply-chain" by providing SMEs with the ability to plan their process innovations as well as improve the design and development of new products. It would connect firms to the existing network of publicly funded R&D resources that are meant (by legislative intent and design) as access points for SMEs to gain technical assistance and information about advanced manufacturing processes.

The National Advanced Manufacturing Portal would provide a current catalog of information about 1) what can be accessed (the portfolios of the cooperative research centers) and 2) what technical assistance and resources SMEs want to access (their inquiries). It would also allow state and local science and technology policymakers to quickly find what federally funded resources are available. It would also allow researchers to determine the relative coverage of S&T resources in a given area or in a targeted technology. This could lead to vastly more efficient S&T policy investment and coordination.

Second, this "open web portal" would be designed to provide an even playing field for firms and allow them to cross technology, industry, and sector boundaries. This open format would not require targeting (and the constant reevaluation of targets) of technology/sector investments. The real-time innovation on the R&D side (the institutions) would determine how they write/update their searchable catalog portfolios and the real-time production (and preproduction) needs of firms define their search queries. No intermediary would be required to guess the scope and definitions of firm needs or technological capacities. No central authority would be required to keep up with the innovations at cooperative research centers.

Almost all private sector or pilot "web-portal" projects are member-based (meaning log-ins or memberships are required) and/or sectoral in coverage. These also typically try to connect up a supply-chain. Also, private sector web portals and databases are largely intended to sell something. These resources either connect suppliers to contractors or sell R&D services, intellectual property, or technical assistance to production firms. This web portal would serve solely as an information intermediary rather than a market intermediary.

The up-front resource requirements to launch and maintain the Portal are comparatively small; cooperative research centers would provide and update the information on their own facilities through a web reporting interface using a standardized format. This reporting would produce the content portion of the National Advanced Manufacturing Portal. The web portal itself would need to be hosted and maintained by an appropriate federal agency.

Design of the Portal

The portal will be in the form of a searchable catalog of publicly funded cooperative research centers. Initial implementation is limited to peer-reviewed facilities (i.e. grant-recipients of public funding) in order to ensure quality of facilities listed. The intended user group is SMEs, but its use is not restricted.

Searchable Fields in the Catalog (drop-down menus):

- Where is the facility? (city, state)
- What is the facility? (name, network, partners)
- Technology specializations (organic photonics, thin film coatings)
- Industry specializations (OPVs, medical devices, printed electronics)
- Sector specializations (energy, health, aerospace, space, defense, consumer electronics)
- Keyword (nuclear, oak ridge, Atlanta)
- What equipment is available to firms? (particle accelerator)
- Is technical assistance available? (Y/N)
- Is training available? (Y/N)
- Is there a fee for access? Different fee structures? (Y/N)
- Are scale-up facilities on site? (Y/N)
- Is there a proposal review process required? (Y/N)
- Are certification services available (environmental, ISO, CCC's)? (Y/N)
- Can small production runs be conducted? (Y/N)

Cooperative Research Center Portfolios in the National Advanced Manufacturing Portal (a brief description written by each CRC elaborating on the following):

- What are the costs/fees for access to equipment?
- What technical assistance is available to firms?
- What are the IP arrangements including non-disclosure agreements?
- What kind of incumbent worker training is available?
- What are the requirements for access?

- What is the training process?
- What is the time line for access?
- What certifications are available?
- Whom do I contact?

Content Generation and Maintenance

Publicly funded agencies would submit and regularly update content on their facilities and resources (as a reporting requirement of public funding). The resulting track-able inquires (counts of inquires on particular technologies, resources, types of centers, equipment) can feed back into a better understanding of the needs of firms, creating the possibility of improved targeting of specific actors and the services most needed (for example: SMEs and assistance with process innovations).

Suggestions for Implementation/Launch:

- Use mailing lists of existing programs to broadcast availability (for example, the MEPs have over 30,000 client firms)
- NIST/MEP might serve as administrative host agency
- Link to manufacturing.data.gov and/or other non-profit and public web portal networking initiatives (such as Autoharvest.com)
- Coordinate with portal initiatives in other Federal agencies that are focused on other aspects of the pre-production process
- The developer of the Portal should be well versed in the variation in cooperative research center structures (university-based, public sector, and public/private partnerships)

CONCLUSION

To enable the United States to successfully translate discoveries into products or applications in manufacturing, the AMP Steering Committee Shared Infrastructure and Facilities Workstream recommends the establishment of a national network of Manufacturing Innovation Institutes (MIIs) to bridge the gap between basic research performed at universities and national laboratories and the work done at U.S. production enterprises. These institutes would serve as an anchor for technology development, education, and workforce training.

MIIs should support prioritized manufacturing technology areas, focusing initially on those recommended by the Advanced Manufacturing Partnership Steering Committee, and subsequently on high priority new technologies as they arise. Future areas of support would be expanded to include emerging technologies that have the greatest potential for commercialization. These areas should be identified using a permanent model and roadmap process for prioritizing investment in advanced manufacturing technology. An open, competitive process with peer reviews should be used to establish the MIIs. (See Annex 1 for a recommended approach to develop this permanent model and roadmap process).

We recommend that at least 5 such Institutes be established in 2012, increasing by 5 per year with the goal of establishing a total of 30 over a 6-year period.

We also recommend the establishment of a National Advanced Manufacturing Portal that would support rapid technology development and commercialization among the SME community by providing a roadmap to existing shared facilities and resources to support their work.

REFERENCES

[1] E. Mills and J. Livingston, "Traversing the Valley of Death," *Forbes.Com*, November 17, 2005.

[2] President's Council of Advisors on Science and Technology, "Report to the President on Ensuring American Leadership in Advanced Manufacturing," June 2011.

[3] National Science Board, "Chapter 4, R&D: National Trends and International Comparisons," in *Science and Engineering Indicators 2010.*

[4] Battelle and *R&D Magazine*, "2012 Global R&D Funding Forecast," December 2011, http://www.battelle.org/aboutus/rd/2012.pdf.

[5] B. Feiereisen, "High Performance Computing for Manufacturing—Why is it Not Used Everywhere?" in *Digital Manufacturing Reports*, October 11, 2011, http://www. digitalmanufacturingreport.com/dmr/2011-10- 11/ high performance computing for manufacturing %E2%80%93 why is it not used everywhere .html.

[6] A. S. Kazi, T. Ristimaki, O. Balkan, M. Kürümlüoglu, J. Finger, and T. Sustar, " Model-Based Collaborative Virtual Engineering in the Textile Machinery Industry: Living Lab Case Study," ICE 2009, 15th International Conference on Concurrent Enterprising, http://85.236. 55.73/Projects/444/ICE%202009%20Conference%20 Paper/3ICE2009 CoVE S Living%20Lab%20Case Balkan.pdf.

[7] Council on Competitiveness, "U.S. Manufacturing—Global Leadership through Modeling and Simulation," 4 March 2009, http://www. compete.org/images/uploads/ File/PDF%20Files/HPC%20Global%20Leadership %20030509.pdf.

In: Advanced Manufacturing in America
Editors: Brandon Cullerne and Michael Laston
ISBN: 978-1-62257-932-7

Chapter 5

CAPTURING A DOMESTIC COMPETITIVE ADVANTAGE IN ADVANCED MANUFACTURING. *ANNEX 3: EDUCATION AND WORKFORCE DEVELOPMENT WORKSTREAM REPORT*[*]

President's Council of Advisors on Science and Technology

ABSTRACT

In June 2011, the President established the Advanced Manufacturing Partnership (AMP), which is led by a Steering Committee that operates within the framework of the President's Council of Advisors on Science and Technology.

In July 2012, the AMP Steering Committee delivered its report to PCAST, entitled *Capturing Domestic Competitive Advantage in Advanced Manufacturing*.

PCAST adopted this report and submitted it to the President. The Steering Committee's report draws on preliminary reports prepared by several "workstreams." These workstream reports have been made available as on-line annexes to the Steering Committee report.

EXECUTIVE SUMMARY

For the United States to remain a competitive force on the world stage, talented employees who have a high level of technical skill are needed to revitalize, sustain, and improve U.S. manufacturing. Unfortunately, the image of manufacturing and the public perception that it can provide long-term, desirable careers have been tarnished.

This negative image is driving the most talented, technically skilled students away from manufacturing to other career paths and creating a deficiency in the quantity and quality of

---

[*] This is an edited, reformatted and augmented version of President's Council of Advisors on Science and Technology, dated July 2012.

the current and future workforce. To attract a robust and highly skilled workforce, the image of manufacturing must change from offering low job security and dull, dirty, and dangerous work to being exciting, engaging, essential, and environmentally sustainable. The same cohesive message from the government, educational institutions, and private industry is needed to change this perception. An equally important need is for some modification to traditional teaching methods used to train the manufacturing workforce at all levels of education. Success in advanced manufacturing and entrepreneurship will require a workforce with fundamental science, technology, engineering, and math (STEM) skills and broad problem-solving skills, decision-making skills, and people skills that do not emerge from a conventional K–12 education. We encourage adoption of Project-Based Learning (PBL) methods in upper K–12 and in community college programs in manufacturing, with some projects selected for their relevance to manufacturing-relevant skills, such as supply-chain management, design for manufacturability, estimation of tolerances and requirements, economics, and team-management. To stimulate these new educational initiatives, educational partnerships between industry, academia, and local and regional governments must be established. A successful rebirth of manufacturing will be ensured only by addressing the underlying structural challenges.

Charge to the Workstream

The Education and Workforce Development (E&WD) Workstream's task was to identify tangible actions that support the availability of a robust supply of talented individuals to provide human capital to advanced manufacturing companies in the United States today and in the future.

The discussions about advanced manufacturing have raised awareness that efforts aimed at creating jobs and opportunities must be matched by initiatives to expand and improve the capabilities of the workforce. Industry participants have consistently raised concerns about the "largest gap" in our educational system—it does not produce people with the technical knowledge, basic business skills, people skills, and problem solving abilities necessary to succeed in a modern manufacturing facility. The Advanced Manufacturing Partnership Steering Committee (AMP SC) must begin efforts to fill this gap if all other aspects of advanced manufacturing are to take root in the United States. We decided to focus much of our initial attention on initiatives that can impact manufacturing education at the community college level because we believe the greatest near-term impact on the workforce can happen at this level.

Process Followed

The E&WD Workstream comprised individuals from six industrial companies, four universities, five Federal agencies, and one community college. The participation of each proved valuable in providing a wide perspective on workforce needs. Weekly teleconferences were held to foster continuity and ongoing dialog. Guest speakers participated in many teleconferences and expanded the knowledge of the workstream. Many of the guest speakers volunteered to work with the AMP to assist with the implementation of recommendations. This interaction proved to be an important part of the process. When the input of a guest speaker was needed beyond what was provided during a teleconference, additional conferences were held.

The topic of education is very broad and presented challenges for the workstream. Early meetings were filled with valuable discussion and provided members a good background on the topic. However, because of the breadth of the topic, getting traction was difficult.

The first action of the workstream was to draft and send a survey to a group of industrial firms soliciting their response to five questions. These responses were used to focus the workstream discussion on community colleges, which was identified as the area that could have the largest impact on closing the workforce's widest educational gap. Survey results are shown in the Supporting Materials section of this report.

After three months of working on the work-life and work-skills gaps of employees entering the workforce for the first time, the workstream expanded the topics of study. Six subgroups were formed, nicknamed workcreeks. Each workcreek focused on one topic of interest to the E&WD Workstream. The areas of focus were Enhancing the Image of Manufacturing, Veterans, Federal Programs, Manufacturing Programs at Research Universities, Standards and Certification, and Attributes of Successful Partnerships. Workcreek leaders were assigned, and the workcreek managed its work independently and reported its findings during a weekly teleconference. This division of labor proved essential and allowed significant progress in a short period of time.

Members read and shared reports from groups such as the National Association of Manufacturing (NAM), the President's Export Council (PEC), Jobs Council, the Manufacturers Alliance for Productivity and Innovation (MAPI), the Information Technology and Innovation Foundation (ITIF), and the U.S. Chamber of Commerce (USCC) and discussed related articles published in trade journal and alumni magazines.

Co-chairs and workstream members attended the four outreach meetings and participated in the E&WD breakout sessions. The discussion and feedback from the attendees confirmed much of our work and gave us confidence that we were on the right track. In each outreach meeting, the attendees also brought new ideas, which became important findings and helped form and harden our recommendations.

In the final weeks before the submittal of the workstream report, each workcreek leader produced an extensive report with his/her findings and recommendations. Each report was a standalone product that could have been submitted as a final report. The next step was to gather the members for a day-long work session. The task was to edit and finalize workcreek reports and begin the processes of consolidating the work into a final report with integrated, powerful, and actionable findings.

In summary, participation was excellent and was vital to the performance of this workstream. The findings are important, and the recommendations are extensive. If the recommendations are adopted, they will have a significant, immediate, and long-lasting positive impact on U.S. advanced manufacturing.

KEY FINDINGS

Summary

Even with unemployment near 9%, advanced manufacturing positions are available in a wide range of industries. These employment opportunities would provide rewarding careers (e.g., in businesses whose products improve energy efficiency and sustain the environment).

Over the past decades, manufacturing jobs have changed. These jobs now require highly skilled workers instead of laborers. The largest gap between manufacturing's needs and new employee skills exists for technicians and equipment operators. This gap has left many workers unqualified for available positions. Community colleges provide some of the missing education, but a significant gap remains. Most of the gap can be found at the secondary level, where many students are not prepared to join the manufacturing workforce. Workers lack general workplace skills such as problem solving, social interaction, and teamwork. Basic communication skills of reading, writing, and mathematics are also inadequate. As a result, businesses often must train employees in areas of STEM before they can make needed contributions. An accreditation system that focuses on the skill needs of the workforce is needed.

This team has focused on initiatives that can expand the workforce in the near term. Findings of greatest importance are as follows:

1. Advanced manufacturing is not confined to emerging technologies. It comprises efficient, productive, tightly controlled processes across a wide spectrum of globally competitive U.S. manufacturers. We define advanced manufacturing as industries where the co-location of manufacturing and design leads to innovation.
2. The quality of the U.S. workforce is not the primary reason manufacturing companies locate facilities outside the United States; however, work-force quality is a concern to industry and a top consideration when making a decision about where a new facility will be located.
3. The largest new-hire skills gaps are for operators and technicians of automated equipment and processes and for trade skills such as welders and electronic technicians. The skills gap spans work-life and job skills.
4. Service members exiting the military possess many of the needed technical and work-life skills.
5. Community colleges grew after World War II to train the returning GIs to join workforce. This founding principle can serve today's need to train returning veterans to meet the needs of advanced manufacturing.
6. The loss of manufacturing jobs over the past 25 years and the negative image of manufacturing careers have driven talented people away from manufacturing careers. Fewer manufacturing jobs result in fewer students being interested in manufacturing, which, in turn, results fewer manufacturing-based courses and degrees being offered by educational institutions. All of these trends are working against the needs of industry and must be reversed.

7. Educational programs that base curricula on project-based experiences seem to be the best at producing graduates who have the skills desired by employers in advanced manufacturing.

Image of Manufacturing

Manufacturing careers are viewed with disdain and skepticism. Reductions in force (RIFs) and offshoring are regularly reported by the media. Repetitive reporting has created an image that manufacturing jobs do not offer job security. Manufacturing jobs within a facility can decline as productivity improves and U.S. businesses outsource or open facilities outside the United States. This trend is likely to continue as companies work to be competitive in the global marketplace. Conversely, one of the highest concentrations of community and individual wealth creation is manufacturing.

The conventional wisdom about manufacturing evokes images of the past and leads one to believe that jobs in this sector are dirty, noisy, repetitive, and dangerous and that manufacturing operations are harmful to the environment.

Manufacturing and "factory work" are denigrated by influential members of society, across all sectors, as a job (not a career) that should be avoided or surpassed through better education.

Jobs for unskilled labor are declining, but jobs for skilled operators and technicians are increasing at a rate that exceeds the availability of qualified candidates. These highly-skilled, creative and innovative professionals are essential to a corporation's long-term competitiveness. In a recent survey conducted by Deloitte and the Manufacturing Institute, manufacturing is nationally viewed as core to our economic prosperity and preferred as an industry for creating local employment. Related to available skills to support manufacturing growth, 82% of manufacturers reported moderate-to-serious gaps in the availability of skilled manufacturing candidates. In addition, 74% of manufacturers report that this skills gap has negatively impacted their company's ability to expand operations. This skills gap has resulted in 5% of all manufacturing jobs going unfilled—even in the face of our current unemployment levels. The focus needs to be on developing a strong pipeline of prepared manufacturing candidates as a key enabler to advancing manufacturing in the United States.

Veterans

Veterans demonstrate many of the work-place skills that are in great demand in advanced manufacturing. These skills include maturity, discipline, the ability to work effectively in a group, and leadership. In addition, many veterans have undertaken extensive technical training that has resulted in skills that could easily translate to manufacturing positions, such as technicians, operators of complex equipment, or craftsmen. Yet, the veteran population is experiencing a higher rate of unemployment than their civilian counterparts. In examining statistics for 2011, the Bureau of Labor Statistics (BLS) found that the unemployment rate for veterans who served in the military at any time since September 2001 (called Gulf War-era II veterans) was 12.1%. The jobless rate for veterans of all eras combined was 8.3%, compared with 8.7% for non-veterans.[1]

In a recent survey conducted by the Manufacturing Institute, 67% of respondents reported a moderate-to-severe shortage of available, qualified workers, and 56% anticipated the shortage to grow worse in the next 3 to 5 years. In addition, the survey found that 5% of current jobs at respondent manufacturers are unfilled due to a lack of qualified candidates.[2] The manufacturing sector recognizes that veterans could provide a robust employee pool, as exemplified in the Manufacturing Institute's new Pipeline Initiative to "connect transitioning military men and women to manufacturing employment through 'high-tech' regional and local career expos."[3] As a result, we recommend actions to:

- Facilitate the matching of skilled veterans with manufacturers, including clear translation of military training and certifications to civilian training and certifications;
- Expand the recognition across the manufacturing sector of the unique skill sets offered by veterans; and
- Ensure that veterans are aware of opportunities for careers in manufacturing (an outcome tightly coupled with improving the image of advanced manufacturing).

Project-Based Learning

Many educational "experiments" are underway in which all or parts of a curriculum are designed around project-based instruction. "New Tech High Schools" have their entire curriculum based on student teams executing industry-specified project challenges. Community colleges often rely on project-based experiences in some classes to provide skills. Some colleges (e.g., The Claremont Colleges in California) base entire years of their curriculum on team projects. Bachelors of Science in Mechanical Engineering (BSME) and other graduate programs at all top universities include team projects in "Capstone Design" courses or other focused single-term or multi-term courses. The emergence of these methods at all levels has created an awareness that these experiences lead to the development of important skills such as decision-making and leadership skills, which go beyond the STEM core. Participants from the industrial sector have emphasized the unique value of these experiences in the education of their best employees. Manufacturers who go into the classrooms provide real-world projects and research opportunities. They support the adoption of project-based learning (PBL) and the revitalization of job shadowing, internships, and apprenticeships. These manufacturer-led partnerships and initiatives have been successful at the local and regional levels in producing graduates who have skills that manufacturers value. The best practices and key attributes of these successful partnerships should be captured and propagated. Governmental entities can encourage these partnerships by funding progressive benchmark initiatives and defunding legacy status quo programs.

Community Colleges

Community colleges grew after World War II to train the returning GIs to join workforce. This founding principle can serve today's need to train returning veterans to meet the needs of advanced manufacturing. The approximately 1,500 community colleges located across the

United States should develop location-specific curricula to meet the needs of local and regional manufacturers.

A successful implementation model currently exists with the National Science Foundation (NSF) Advanced Technological Education (ATE) Program. This program emphasizes the role of community colleges as the main providers of technician education in the United States. ATE centers and projects at community colleges, in partnership with universities, secondary schools, business and industry, and government agencies, design and carry out model work-force development initiatives.

Technology funding managed by the NSF, the Department of Defense (DOD), the Department of Energy (DOE), the Department of Commerce (DOC), and other agencies includes investments in technologies for advanced manufacturing. While this funding is significant, it is overwhelmingly delivered to universities, national labs, and industry and has almost no impact on the community colleges' educational programs that are critical for training the next-generation advanced manufacturing workforce.

Colleges and Universities

A survey of the six universities showed that none has a named undergraduate degree in manufacturing; however, several have at least one comprehensive program at the graduate level. Many baccalaureate-level engineering degree programs in 4-year colleges and universities have very little in the way of manufacturing science and technology content in their curricula. The lack of exposure to the engineering and systems aspects of manufacturing is a key factor in the low level of interest that program graduates have in pursuing a career and/or further studies in advanced manufacturing. Major research universities must play a key role in defining the fundamental elements of the discipline of advanced manufacturing and in producing the next generation of educators and industrial leaders. In so doing, these institutions will not only add to the profession, but will also greatly improve the image of manufacturing as a challenging and rewarding career.

This effort should be aimed at programs and degrees that give the student a comprehensive view of manufacturing and that provide a technological and an operational perspective to the student in a professional engineering context.[4] If one starts with the premise that U.S. manufacturing excellence includes the need for graduates from such degree programs,[5] a review of current programs reveals that we are not addressing these needs nationally. What emerges is a picture of local programs that rise and fall with local enthusiasm and industry interest and that are, on the whole, isolated and independent. Likewise, universities have not learned where manufacturing best fits in academia. It does not fit well into normal boundaries of degree programs, departments, or even schools and, as a result is often marginalized. Also, typical research university interactions with industry are with R&D and not manufacturing organizations.

Certifications and Accreditations

An efficient market for employees who have needed knowledge and skills depends on reliable and appropriate credentials and/or certifications. To succeed, any new assessments, accreditations, and credentials require a critical mass of national recognition and acceptance and adoption by industry, education, and government. Such certifications work when they

- Involve quality assessments, accurately gauging worker skills;
- Include an accreditation regimen that ensures program quality and alignment with the changing needs of industry; and
- Result in nationally portable, industry-recognized support, preferential consideration, and job search mobility.

Public/Private Partnerships

Traditional approaches to technical education are mismatched to the needs of advanced manufacturing and have contributed to unfilled jobs at a time of high unemployment. Regional programs and partnerships successfully close the skills gap.

There are many examples of successful industry-academia-government partnerships that have improved the education system, but no two partnerships alike, even in the same region with similar partners.

Fundamentally, the partnerships that have had best outcomes have been built on six important pillars: (1) partners have a passion for learning and a vision for the future, (2) partners embrace the case for change, (3) convening organizations share their expertise, (4) collaboration, committed involvement, and sense of community combine for success, (5) the partners' specific roles are clear, and (6) stakeholders remain flexible to meet manufacturers' needs.

When industry partners work with academia, all parties benefit. Manufacturers that engage at the K–12 level affect students across society, regardless of the career path they choose. These partnerships help ensure that students are provided with relevant teaching that will enable them to contribute to society and provide for themselves upon graduation.

RECOMMENDATIONS

Summary

These findings require a call to action from government, academia, and industry. Our recommendations include the following:

1. The image of manufacturing needs a complete restoration, removing false impressions based upon partial truths and realities of the past. The Advanced Manufacturing NPO should launch a nationwide Ad Council[6] campaign to restore the image of manufacturing careers, with outreach support from existing associations such as the Society of Manufacturing Engineers (SME), the National Association of Manufacturers (NAM, and the Institute of Industrial Engineering (IIE).
2. Veterans possess many of the missing skills that are crucial to advanced manufacturing. Veterans must be aware of career opportunities in manufacturing, and

industry must be aware that veterans can provide solutions to many existing employment problems. This solution is truly elegant. Veterans win, business wins, and our nation is made stronger. We recommend providing tuition aid for returning veterans who enroll in manufacturing related programs in community colleges and universities (e.g., through the post 9-11 GI Bill) and cost sharing by manufacturing companies interested in hiring and retraining veterans. We support tax credits for employers that hire and train veterans (e.g., those provided in the Vows To Hire Heroes Act). Community colleges near military bases where returning veterans are likely to locate after return should receive special assistance to create and support programs to retrain veterans. A "GI Bill" for retraining and employing veterans should be initiated to carry and support all of these initiatives.

3. Community colleges already enroll many of the people who should train for advanced manufacturing. They have partnerships, infrastructure, and teaching methods that focus on regional needs. The Nation needs to invest in improvements in community colleges and promote engagement among community colleges, and industry, universities, national labs, and K–12 programs. Modest changes to agency solicitations to encourage partnering with community colleges could instantly create stronger regional community college partnerships with the industry, universities and national labs that routinely seek agency R&D funding.
4. Certifications and accreditations for skills in advanced manufacturing are needed. These educational certificates should be portable from institution to institution (enabling mobility between regions) and from colleges to jobs and back. Nationwide associations, such as NAM and the MI, should initiate and coordinate a register of certifications that are available in all regions and that can be "stacked" one after another to assemble complete programs of training in advanced manufacturing.
5. The importance of manufacturing-related content in university education needs to be highlighted. The Accreditation Board for Engineering and Technology (ABET) can serve in this role by imposing modest changes to the "attributes and objectives" of accredited undergraduate programs in engineering and by encouraging the insertion and use of manufacturing-inspired challenges in the undergraduate curriculum.
6. Federal agencies are already sponsors of advanced manufacturing in many technology programs. Modest changes to the review and selection criteria of government agency programs could lead to a much greater impact on the development and commercialization of technologies for advanced manufacturing and on the creation of opportunities for attracting and training new members of the advanced manufacturing workforce.
7. The creation of graduate degree programs in advanced manufacturing (e.g., through the NSF Integrative Graduate Education and Research Traineeship (IGERT) program and national MS and PhD Fellowships in advanced manufacturing) should be encouraged.

The education and work-life skills gaps are problems faced by large and small manufacturers throughout our country. Left unaddressed, these problems will continue. The good news is the solutions are known, discrete, and solvable with a sustained long-term approach. The following subsections support each of these recommendations with more detailed discussion.

Image of Manufacturing

Engage the Ad Council to develop and launch a multi-year (3-year minimum) Public Service Announcement (PSA) awareness campaign to improve the image of manufacturing. The Advanced Manufacturing NPO would lead the campaign under the guidance of the industrial members of the Ad Council's Board of Directors.

- Peers have a significant—perhaps the most significant—effect on teenage students. Focus the PSA campaign on high school and community college students. A peripheral benefit would be that parents' and educators' image of manufacturing would also improve.
- The PSA campaign is projected to have an out-of-pocket cost of $800,000 to $1,000,000 annually. This could be funded by the DOC's Advanced Manufacturing NPO and through industry and professional society foundations.
- Pro bono costs for creative development and media placement range for $25 to $30 million for a typical PSA campaign.
- Consider modeling the program on Army of One, Got Milk, or Essential 2 advertisements
- Consider a message that speaks to patriotism, creativity, innovation, potential, and the critical link that manufacturing will play in America's future.
- Manufacturing careers need to be elevated so they are viewed as professions and not just as jobs.
- A subset of the messages should be regional in nature, highlighting individual sectors that are important regionally (Robotics in Automation Alley [Michigan]), aerospace (South Carolina), advanced electronics (California). These messages could affect students of all ages.

Veterans

Design and develop an educational advanced manufacturing module that emphasizes opportunities, challenges, and benefits of manufacturing to individuals and society. This module would include the types of businesses, examples of processes and equipment, and the skills of people who work in that environment.

- This module should be inserted into DOD's Transition Assistance Program (TAP) at the highest organizational level to mitigate the local variability in administration of the program but should also be available before and after transition.
- The module would be funded, directed, and managed by the DOD.
- The Ad Council would take the advanced manufacturing TAP module and revise it for civilian use. The module would be made available free to professional societies such as SME, NAM and IIE and to educational institutions. The military and civilian modules would be produced as standalone presentations that would not require the presenters to have any knowledge of advanced manufacturing.

Veterans have many of the work-life and jobs skills in high demand but missing in the general workforce. The Department of Labor (DOL) and the DOD should also focus on the design and development of job-seeking tools that facilitate the translation of military occupational codes (e.g., U.S. Army Military Occupational Specialties (MOS)) to civilian occupational credentials (e.g., MI's Manufacturing Skills Certification System). DOL and DOD should consider endorsing the MI's Pipeline program as the go-to point for veterans interested in manufacturing and for manufacturers interested in veterans. DOL and DOD should also encourage manufacturers to commit to using such tools and local and regional outreach (e.g., job fairs and partnership with State Veterans Affairs departments) to reach veterans in post-transition.

Community Colleges

Community colleges grew after World War II to train the returning GIs to join workforce. This founding principle can serve today's need to train returning veterans to meet the needs of advanced manufacturing.

- The approximately 1,500 community colleges located across the United States should develop location-specific curricula to meet the needs of local and regional manufacturers.
- The Advanced Manufacturing NPO should arrange a meeting of community colleges in Washington D.C. The American Association of Community Colleges (AACC) represents over 1,100 community colleges across the United States and would be a likely source to help coordinate the meeting. The purpose of this meeting would be to explain the need, solicit long-term support, and roll out the advanced manufacturing module for civilian use. Presentation would also be given by participants in this workstream.
- Community college curricula and credentials should be developed and standardized to meet the needs of manufacturing.

Universities

High schools, community colleges and universities must work together to educate our workforce to meet the needs of advanced manufacturing environments. Universities should take the lead in linking and coordinating the "education supply chain." The system should be a pull system that reaches out to regional industries and connects their needs with educational institutions. This effort could become part of the role of Manufacturing Innovation Institutes (MII).

- Engineers with varying educations are required for advanced manufacturing environments. Manufacturing engineering courses needs to continue to be part of or be added to Bachelor of Science engineering programs, as recommended elsewhere in this Annex.

- Research universities should establish new Masters-level professional degrees in Manufacturing Leadership. The major research universities have a special responsibility to establish what advanced manufacturing means and why it is so vitally important. These degree programs should be comprehensive in their integration of technologies (e.g., robotics and advanced automation) with methods (e.g., supply-chain management). Once established, the major universities must collaborate to establish a new educational model and uniform standards that can propagate nationally to an aligned set of education and training programs at the secondary, certificate, sub-baccalaureate, and baccalaureate levels.
- Industry associations, professional societies and educational organizations such as the IIE, SME, NAM, American Chemistry Council (ACC), and the USCC should form a coalition that will
  - Establish a path forward toward a national framework of standards, accreditations, and certifications at each level of the manufacturing workforce
  - Work directly with high schools and community colleges within their communities to stress the importance of using the advanced manufacturing module.

The Federal Government's help is needed to attract students and faculty to new manufacturing degree programs (e.g., through the introduction of National Manufacturing Fellowships, Veterans Leadership in Manufacturing Fellowships, funded "Traineeships," and curriculum/program development funds). For example, programs that would extend the NSF IGERT program to focus on MS-level manufacturing degrees or the inclusion of advanced manufacturing in the Department of Education's Graduate Assistance in Areas of National Need (GAANN) program would spur university activity. Potential action items include:

- Create "Presidential Fellowships and Scholarships" to encourage U.S. students interested in manufacturing careers to pursue professional degrees in this area.
- Modify the Department of Education's GAANN program to have a focused solicitation on manufacturing fellowships/scholarships at the university and community college levels. Structure this modified program to encourage collaboration between industry, community colleges, and universities or have separate scholarship programs aimed at the different educational levels.

Accreditations and Certifications

Ensure this path forward incorporates two key functions (accreditation and certification) and results in two key outcomes (common education and training standards) to satisfy current and emerging competencies and portable certifications for individuals. Leverage and/or expand existing models that have proven successful.

- **Accreditation.** ABET validates applied science, computing, engineering, and technology education programs by using an agreed-upon set of criteria to assess the quality of degreed programs at the associate, bachelors, and graduate levels for engineering and engineer technology.[7]

- **Certification.** The Manufacturing Institute (MI) uses the DOL Advanced Manufacturing Competency Manufacturing framework[8] to align existing credentials to a national certification system and uses this framework to assess future and incumbent workers' knowledge and skills. The MI has also established an extensive coalition of partnering organizations that provide certifications for most of the competencies outlined in the DOL model. Current partners include ACT, the Manufacturing Standards Skills Council (MSSC), the American Welding Society (AWS), the National Institute for Metalworking Skills (NIMS), and the SME.

CONCLUSION

While it is impossible to separate the education system into discrete pieces, the AMP E&WD workstream believes that the most impactful recommendations for improvement are at the high school and community college levels, followed by undergraduate education. Industry has identified content-mastery and soft-skills deficiencies that could be impacted by changing the way students are taught through the use of project-based learning, which has proven its unique capacity to prepare young people for a 21st century workplace. The content-mastery and soft-skills deficiencies can be remedied through this approach when coupled with a marked improvement in teaching skills.

To further secure and develop our talent pipeline, we believe the following must be accomplished through public/private partnerships.

- Create an aggressive, integrated "Image of Manufacturing" public service announcement campaign that would raise awareness and correct misperceptions of manufacturing in the United States;
- Provide support for veterans who possess the skills to fill technical manufacturing jobs, starting with the addition of a training module on advanced manufacturing to TAP materials;
- Build a certification and accreditation program to create national standards for advanced manufacturing;
- Enhance the role of research universities in defining the discipline of advanced manufacturing;
- Invest in community colleges and project-based curricula to build advanced manufacturing skills.
- Create a coordinated interagency initiative under the direction of the Advanced Manufacturing NPO to expand existing programs that enable students, faculty, and post-doctoral researchers to interact directly with manufacturers through the establishment of a high-profile public/private national internship and fellowship project; and
- Leverage programs such as the NSF ATE to develop advanced manufacturing education programs.

SUPPORTING MATERIALS EDUCATION AND WORKFORCE DEVELOPMENT WORKSTREAM

Enhancing the Image of Manufacturing

For the United States to remain a competitive force on the world stage, talented employees who have a high level of technical skill are needed to revitalize, sustain, and improve U.S. manufacturing. Unfortunately, the image of manufacturing and the public perception that it can provide a long-term, desirable career have been tarnished. This negative image is driving the most talented technically skilled students away from manufacturing to other career paths and is creating a deficiency in the quantity and the quality of the current and future workforce. To attract a robust and highly skilled workforce, the image of manufacturing must change from low job security, dull, dirty, and dangerous to exciting, engaging, essential, and environmentally sustainable. The same cohesive message from government-supported programs, educational institutions, and private industry is needed to change this perception.

Key Findings

- Manufacturing is viewed as lacking job security. RIFs and offshoring are regularly reported by all arms of the media. Repetitive reporting has created an image that manufacturing jobs do not offer job security. Manufacturing jobs within a facility can decline as productivity improves and U.S. businesses outsource or open facilities outside the United States. This trend is likely to continue as companies work to be competitive in the global marketplace. Conversely, one of the highest concentrations of community and individual wealth creation is manufacturing.
- The conventional wisdom about manufacturing evokes images of the past and leads one to believe that jobs in this sector are dirty, noisy, repetitive, and dangerous and that manufacturing operations are harmful to the environment.
- Manufacturing and "factory work" is denigrated by influential members of society, across all sectors, as a job (not a career) that should be avoided or surpassed through better education.
- Jobs for unskilled labor are declining but jobs for skilled operators and technicians are increasing at a rate exceeding the availability of qualified candidates. These highly-skilled, high wage, creative and innovative professionals are essential to a corporation's long term competitiveness.

Recommendations

- Design an "Advancing Manufacturing" campaign to transform the image of manufacturing through the Ad Council similar to An Army of One, Got Milk, or Essential2.
- Appeal to the public's sense of patriotism and the critical link that manufacturing will play in America's future.
- Consider a message that speaks to creativity, innovation, personal potential, and public service.

- A subset of the messages should be regional in nature, highlighting individual sectors that are important regionally (Robotics in Automation Alley (Michigan), aerospace (South Carolina), advanced electronics (California). These messages could impact students of all ages.
- Determine if the Advanced Manufacturing NPO will manage the initiative.
- Collaborate with professional organizations such as IIE, SME, NAM, AACC, and USCC to work directly with high schools and community colleges to spread the messages of the importance of manufacturing.
  - Large trade associations should provide "conversation starting" materials.
  - Professional organizations could commit to send 10% of their membership into classrooms to start the conversation about manufacturing.

Veterans

Veterans demonstrate many of the workplace skills that are in great demand in advanced manufacturing. These skills include maturity, discipline, the ability to work effectively in a group, and leadership. In addition, many veterans have undertaken extensive technical training that has resulted in skills that could easily translate to manufacturing positions, such as technicians, operators of complex equipment, or craftsmen.

Yet, the veteran population is experiencing a higher rate of unemployment than their civilian counterparts. In 2011, the Bureau of Labor Statistics (BLS) found that the unemployment rate for veterans who served in the military at any time since September 2001 (called Gulf War-era II veterans) was 12.1%. The jobless rate for veterans of all eras combined was 8.3%, compared with 8.7% for non-veterans.[9].

At the same time, a recent survey conducted by the MI, 67% of respondents reported a moderate-to-severe shortage of available, qualified workers, and 56% anticipated the shortage to grow worse in the next 3 to 5 years. In addition, the survey found that 5% of current jobs at respondent manufacturers are unfilled due to a lack of qualified candidates. [10] The manufacturing sector recognizes that veterans could provide a robust employee pool, as exemplified in the MI's new Pipeline Initiative to "connect transitioning military men and women to manufacturing employment through 'high-tech' regional and local career expos."[11]

As a result, achieving the following is desirable:

- Facilitate the matching of skilled veterans with manufacturers, including clear translation of military training and certifications to civilian training and certifications;
- Expand the recognition across the manufacturing sector of the unique skill set offered by veterans; and
- Ensure that veterans are aware of opportunities for careers in manufacturing (an outcome tightly coupled with improving the image of advanced manufacturing).

Recommendations: Near term

- Design and develop an educational advanced manufacturing module that emphasizes opportunities, challenges, and benefits of manufacturing to individuals and society. This module would include the types of businesses, examples of processes and equipment, and the skills of people who work in that environment. This module should be inserted into DOD's Transition Assistance Program (TAP) at the highest organizational level to mitigate the local variability in administration of the program but should also be available before and after transition.

- After transition, design and develop job-seeking tools that facilitate the translation of military occupational codes (e.g., U.S. Army MOS) to civilian occupational credentials (e.g., MI's Manufacturing Skills Certification System). Consider endorsing the MI's Pipeline program as the go-to point for veterans interested in manufacturing and for manufacturers interested in veterans.

Recommendation: Longer term

- Identify key characteristics of existing successful models within industry, trade associations, and veterans organizations that are already focused on Veterans (e.g., Intel's Veterans/Clubhouse Initiative, Northrop Grumman's Operation Impact, The United Association (UA) Union of the Plumbers, Fitters, Welders, and Heating, Ventilation, and Air Conditioning (HVAC) Service Techs Veterans In Piping (UA VIP) program). Successful programs should include a manufacturing-specific focus, information on the image/career potential of advanced manufacturing, available tools to easily translate Military Occupational Classifications (MOCs)/MOSs to civilian certification, and local and regional outreach (e.g., Job Fairs and partnership with State Veterans Affairs departments to reach veterans in post-transition).

Review of Federal Programs

Efficient use of resources in support of initiatives for education in advanced manufacturing must be a priority. Opportunities exist today to leverage changes in current Federal programs that directly and immediately impact the development of the workforce for advanced manufacturing. There are three primary areas of opportunity to promote engagements between universities, community colleges, industry, and national labs that stimulate the use of manufacturing-inspired challenges and provide opportunities for project-based educational programs.

Key Findings

- **The NSF ATE.** A successful implementation model currently exists with the ATE. This program emphasizes the role of community colleges as the main providers of technician education in the United States. ATE centers and projects at community colleges—in partnership with universities, secondary schools, business and industry, and government agencies—design and carry out model workforce development initiatives.
- **Manufacturing content in Bachelor-of-Science-level education.** Many baccalaureate-level engineering degree programs in 4-year colleges and universities have very little in the way of manufacturing science and technology content in their curricula. The lack of exposure to the engineering and systems aspects of manufacturing is a key factor in the low level of interest that program graduates have in pursuing a career and/or further studies in advanced manufacturing.

- **Community college partnerships.** Technology funding managed by NSF, DOD, DOE, DOC, and other agencies includes investments in technologies for advanced manufacturing. While this funding is significant, it is overwhelmingly delivered to universities, national labs, and industry and has almost no impact on the community colleges' educational programs that are critical for training the next-generation advanced manufacturing workforce.

Recommendations

- Expand ATE programs to include broader dissemination of successful projects and centers. Consideration should be given in new center and project proposals consistent with AMP SC recommendations.
- Coordinate development of MIIs in partnership with ATE Centers to leverage and expand existing activities and infrastructure.
- Create opportunities for specific, meaningful, and economically driven engagement between industry, the Advanced Manufacturing NPO, and the ATE program.
- To create broad impact on engineering education and manufacturing workforce development, a two-fold approach is suggested:
  - Create an industry pull for fresh engineering graduates, along with the requisite level of manufacturing knowledge and training.
  - Update the criteria specified by the ABET to require manufacturing content in engineering curricula.
- Change scoring mechanisms, algorithms, and other aspects of Federal solicitations for technologies supporting advanced manufacturing so that they will specifically encourage partnerships with community colleges.

Manufacturing Programs at Research Universities

Major research universities must play a key role in defining the fundamental elements of the discipline of advanced manufacturing and in producing the next generation of educators and industrial leaders. In so doing, these institutions will not only add to the profession, but will also greatly improve the image of manufacturing as a challenging and rewarding career.

This effort is aimed at programs and degrees that give the student a comprehensive view of manufacturing and that provide a technological and an operational perspective to the student in a professional engineering context.[12]

Key Findings

- If one starts with the premise that U.S. manufacturing excellence includes the need for graduates from such degree programs,[13] a review of current programs reveals that universities are not addressing these needs nationally.

  What emerges is a picture of local programs that rise and fall with local enthusiasm and industry interest and that are, on the whole, isolated and independent.

  Likewise, universities have not learned where manufacturing best fits in academia. It does not fit well into normal boundaries of degree programs, departments, or even schools and, as a result is often marginalized. Also, typical research university interactions with industry are with R&D organizations and not manufacturing organizations.

Recommendations

- Research universities should establish new Masters-level professional degrees in Manufacturing Leadership. The major research universities have a special responsibility to establish what advanced manufacturing means and why it is so vitally important. These degree programs should be comprehensive in their integration of technologies (e.g., robotics and advanced automation) with methods (e.g., supply-chain management).
  Once established, the major universities must collaborate to establish a new educational model and uniform standards that can propagate nationally to an aligned set of education and training programs at the secondary, certificate, sub-baccalaureate, and baccalaureate levels.
- The Federal Government's help is needed to attract students and faculty to new manufacturing degree programs (e.g., through the introduction of National Manufacturing Fellowships, Veterans Leadership in Manufacturing Fellowships, funded "Traineeships," and curriculum/program development funds). For example, programs that would extend the NSF IGERT program to focus on MS-level manufacturing degree development or the inclusion of advanced manufacturing in the Department of Education's GAANN program would spur university activity.
- A coalition of companies must participate in the aforementioned recommendations by providing funds, people, and advice. For example, industrial partners should work with university coalitions to facilitate the transition of graduates into manufacturing leadership positions.
  Mid-degree internship programs that expose students to careers in Manufacturing Leadership should be provided. Industrial representatives should serve as mentors and role models to students entering and emerging from these new programs. Industry should participate in the development of curriculum modules and in providing and supporting infrastructure at universities.

Accreditations and Certifications

Manufacturing jobs have changed. These jobs now require highly skilled workers instead of laborers. The largest gap between manufacturing's needs and new employee skills exists in technicians and equipment operators. This gap has left many workers unqualified for available positions. Community colleges provide some of the missing education, but a significant gap remains. Most of the gap can be found at the secondary level, where many students are not prepared to join the manufacturing workforce. Workers lack general work-place skills such as problem solving, social interaction, and teamwork. Basic communication skills of reading, writing, and mathematics are also inadequate. As a result, businesses often must train employees in areas of STEM before they can make needed contributions. An accreditation system that focuses on the skill needs of the workforce is needed.

Key Findings

An efficient market for employees who have needed knowledge and skills depends on reliable and appropriate credentials and/or certifications. To succeed, any new assessments, accreditations, and credentials require a critical mass of national recognition and acceptance and adoption by industry, education, and government. Such certifications work when they

- Involve quality assessments, accurately gauging worker skills;
- Include an accreditation regimen that ensures program quality and alignment with the changing needs of industry; and
- Result in nationally portable, industry-recognized support, preferential consideration, and job search mobility.

Recommendations

- Create a coalition of industry associations, professional societies, and educational organizations to establish a path forward toward a national framework of standards, accreditations, and certifications at each level of the manufacturing workforce.
- Ensure this path forward incorporates two key functions (accreditation and certification) and results in two key outcomes (common education and training standards) to satisfy current and emerging competencies and portable certifications for individuals.
- Leverage and/or expand existing models that have proven successful. Two such programs include
  - *Accreditation.* ABET validates applied science, computing, engineering, and technology education programs by using an agreed-upon set of criteria to assess the quality of degreed programs at the associate, bachelors, and graduate levels for engineering and engineer technology.[14]
  - *Certification.* The MI uses the DOL Advanced Manufacturing Competency Manufacturing framework[15] to align existing credentials to a national certification system and uses this framework to assess future and incumbent workers' knowledge and skills. The MI has also established an extensive coalition of partnering organizations that provide certifications for most of the competencies outlined in the DOL model. Current partners include ACT, MSSC, AWS, NIMS, and SME.
- Skill requirements and credentials require continuous updating to accommodate the changing needs of workers and the manufacturing sector.
- Identify gaps and opportunities in these existing models (or others) and include additional organizations where appropriate.

Attributes of Successful Partnerships

There are many examples of successful industry-academia-government partnerships that have improved the education system, but no two partnerships are exactly alike, even in the same region with similar partners.

Fundamentally, the partnerships that have had best outcomes have been built on six important pillars: (1) partners have a passion for learning and a vision for the future, (2) partners embrace the case for change, (3) convening organizations share their expertise, (4) collaboration, committed involvement, and sense of community combine for success, (5) the partners' specific roles are clear, and (6) stakeholders remain flexible to meet manufacturers' needs.

Key Findings

- When industry partners with academia, all parties benefit. Manufacturers that engage at the K–12 level affect students across society, regardless of the career path they choose.
- These partnerships help ensure that students are provided relevant teaching that will enable them to contribute to society and provide for themselves upon graduation.
- Manufacturers who go into the classrooms provide real-world projects and research opportunities, the adoption of project-based learning, and the revitalization of job shadows, internships, and apprenticeships.
- Governmental entities encourage partnerships by funding progressive benchmark initiatives and defunding legacy status quo programs.

Recommendations

- Create "Presidential Fellowships and Scholarships" to encourage U.S. students interested in manufacturing careers to pursue professional degrees in this area.
- Modify the Department of Education's GAANN program to have a focused solicitation on manufacturing fellowships/scholarships at the university and community college levels. Structure this modified program to encourage collaboration between industry, community colleges, and universities or have separate scholarship programs aimed at the different educational levels.
- Create a national network of manufacturing educators by integrating educational programs among NSF, the Department of Education, and the DOL to share best practices, curricula, and resources.
- Establish a program at NSF that funds the acquisition or development of shared equipment for research and/or education purposes.

Survey of Industrial Firms (with Response Summary) – Used as Data to Support Initial Prioritization

1. What elements of education and/or skills are you seeing now in your candidates that are beneficial to your business?

- Math, science, engineering emphasis, with computer literacy, advanced math, basic science/tech/engineering—especially for advanced technician positions that directly support manufacturing operations.

- Advanced computer skills, more advanced degrees and professional certifications, more candidates with formal post-secondary education, and good general writing, reading, communication, and mathematic skills
- Public high school graduates. Weak math skills and no applicable manufacturing skills, except perhaps for entry level material handling and/or stockroom kitting.
- University Graduates. Meet the minimum skills required for entry-level, general machining job descriptions.
  - Technical. No difficulty in finding well-educated candidates for Manufacturing Engineer, Industrial Engineer, or Materials Engineer roles. Use the Pratt & Whitney Manufacturing Engineer Development (MED) Rotational Program to link academic training with practical floor experience.
  - Non-technical. No difficulty in finding candidates for Supply Management, Finance, and Materials Management roles. At the United technologies Corporation (UTC), the Financial Leadership Program (FLP) rotation program is attracting and developing good talent. In 2010, we initiated rotational programs in Supply Management and Materials Management to bridge the academic to practical experience gap for these disciples, similar to MED. The key for future competitiveness in the United States will be the adoption of life learning and education from the shop floor and up.
- For our professional needs, engineering skills are well taught. For our hourly needs, we wage a constant battle to identify candidates with good math/reading/technical skills.
- For PhD, strong Chemical Engineering and Materials Science and Engineering skills. For Techs, strong hands-on project and equipment know-how.

2. At what educational level do you notice the largest gap or deficiency in skills in new hires (high school, community college, or university)?

- High School: 43% of those who responded
- Community colleges: 67% of those who responded
- Universities: 29% of those who responded
  - With the exception of computer literacy, high school candidates tend to possess few skills. This fact, coupled with generally poor social skills, problem-solving ability, and poor work ethic, make this educational level the one that has the largest deficiency in skills.

3. What are the most significant elements (education or skill sets) missing in your job candidates?

- Education: 43% of those who responded
- Skill Sets: 78% of those who responded
  - Practical application and integration of the courses of study. We end up training people how to solve real problems and do work.
  - Vocational, problem solving, and social skills and work ethic, leadership, and business acumen are the more significant missing elements.
  - Critical thinking, math, and electronics abilities are often deficient.

- For basic math and writing skills, and hands-on project experience, the greatest skill deficiency is found in advanced level technicians/manufacturing operators.

4. Have you located a manufacturing plant(s) outside the United States due, in part or in full, to lack of talented employees? If yes, what skill was missing?

- Yes: 29% of those who responded
- No: 71% of those who responded
  - U.S. corporate tax policies relative to other countries is a much greater factor.
  - Skills missing: basic math and project disciplines; apprentice training.
  - See below in general comments.

5. Are there programs or degrees in local schools (high schools, community colleges, or universities) you believe have "best practice" education programs that meet the needs of advanced manufacturing plants? If yes, please summarize and provide the institutions name and a contact person.

- Yes: 67% of those who responded
- No: 43% of those who responded
  - "Tech Prep" programs. Combination hands-on and academic programs. Project-based learning; apprentice learning teams.

6. Please add your thoughts and comments not addressed by this survey.

- As human interaction in advanced manufacturing processes decreases, the quantity of employee base will also decrease. A critical element at our company is the ability to bring new innovations to market rapidly, and, for this, the skills are much less defined than in the past. The ability to link R&D and manufacturing will be an important element if we wish to pursue a "invented here, made here" approach.
- Educated employees are not limiting our ability to manufacture in the United States (or anywhere else for that matter). Finding the right people takes effort, but it pays off.
- Facilities across the United States have similar views of the strengths and weakness of the education system.
- Gaps by educational level
  - High school (for shop floor). Intermediate math, technical literacy, very little practical shop training with basic tools/gages/simple machines
  - Community college (typically for shop floor). Statistics (for process control), technical literacy, technical writing, limited shop training with basic tools/gages/ simple machines.
  - Military (for shop floor). Often greater technical literacy and better familiarity with basic tools and gages than community college students.
  - University. Very few degrees in manufacturing engineering. Stronger emphasis needed on practical elements of material science for machining and special processes. Difficult to secure work visas for international students to fill key manufacturing engineering roles.

- Gaps by function
    - Shop floor technicians. Intermediate math, statistics, technical literacy, technical writing, practical training with basic tools/gages/simple machines
    - Process planners and Numerical Control Machine (NCM) programmers. Very few young U.S. individuals have these skills. Community college co-op program would be helpful here.
    - Technical: Need for more experienced manufacturing engineers. Stronger emphasis needed on practical elements of material science for machining and special processes. Difficult to secure work visas for international students to fill key manufacturing engineering roles. Opportunity for university-level co-op programs.
    - Professional: Aging manufacturing workforce with experience in production planning/scheduling/logistics/advanced quality principles. Opportunity for university-level co-op programs. Built/expanded international factories in part due to skill gaps in the United States. A new machining factory in Singapore supported by training dollars from Singapore government. Generally stronger intermediate math, statistics, and technical literacy at the high school level in Singapore than in the United States. Acquired a factory in Poland near a technical university with 30,000 engineering students and research programs in advanced machining and fabrications. Process planners and manufacturing engineers more readily available than in the United States
- Educational programs in the United States that have worked for us:
    - Solid success recruiting factory/shop technicians from individuals leaving U.S. military.
    - Successful co-op program with Northeastern University through which we have recruited individuals with good educational background in supply-chain management.
    - Successful UTC Operation Leadership Program in which university graduates rotate through assignments to gain professional experience in shop supervision, quality, and supply chain roles
    - Many candidates have the technical skills but lack motivation. Whether characterized by poor attendance or an "it's good enough" mentality, many seem to think they are owed a job.
    - Aforementioned comments focused on vocational training. There is also a critical need to ensure that adequate chemical engineers are being trained (BS, MS & PhD) to support non-bio based businesses.

Ad Council Proposal

Extensive research and review would be conducted during the campaign's formative stage to inform the most effective communications strategy and creative concepts. This research would include a literature review, expert panel and exploratory qualitative research (either traditional focus groups or one-on-one interviews, in-home groups or interviews, or consumer ethnographies). Typically, the Ad Council travels to at least three markets for a given campaign and speaks with a broad range of consumers within the target audience.

After the campaign's creative concepts have been developed, they will be tested qualitatively or quantitatively with consumers to gauge their response to the advertising. This will help determine if the ads will be effective in gaining the target audience's attention, conveying the main message, and persuading the audience to take action.

The campaign's success will be measured in three ways:

- **Media measurement.** The campaign's donated media support will be monitored to estimate the number of ad placements, media impressions generated, and the monetary value of each of these placements.
- **Consumer response.** All forms of consumer response will be monitored, including website traffic, brochure requests/downloads (if appropriate), email sign-ups, etc.
- **Tracking survey.** A national benchmark survey of the target audience will be conducted before the release of the campaign, followed by annual post-wave surveys. The tracking study will gauge trends over time among the target audience. Measures include awareness of the issue, recognition of the advertising, relevant attitudes and relevant behaviors.

Based on this research, the Ad Council would develop creative concepts that are appropriate for motivating the target audience. The campaign will be peer reviewed and approved at three critical stages: campaign strategy, creative concepts, and final advertising materials. This process will be conducted by the Ad Council's Campaign Review Committee (CRC), which is made up of a panel of the nation's top advertising executives who meet monthly and provide feedback for all of the Ad Council's campaigns.

The Ad Council, tapping into the vast pro bono resources of its media partners, would distribute the campaign to its unique network of 33,000 media outlets nationwide (TV, radio, print, out-of-home, and Internet). Ad Council campaigns are also distributed by our Regional Managing Directors, who are located in the country's top Direct Marketing Associations (DMAs) and cover 86% of the country's media markets.

In addition, the Ad Council's National Accounts team maintains relationships with the nation's leading media companies on a corporate and national level to secure top-level media commitments. In addition, its Media Marketing team engages the support of trade and industry associations and coordinates the distribution of each campaign, using unique communication strategies and our staple of general media contacts.

- On average, each general market Ad Council campaign garners $25 to $30 million in donated media. (More broadly, the Ad Council secures about $1.4 billion dollars in donated media for its entire docket.)
- Beyond traditional public-service advertising, the Ad Council also relies on innovative communications, using new communications tools such as bus shelters, yellow pages, taxi cab tops, in-school programming, and cinema advertising and other emerging media outlets (e.g., personal data assistants (PDAs), video email, satellite radio and interactive television).

If funding permits, the Ad Council, in partnership with a reputable public relations agency, could help to create a fully integrated Public Relations (PR) program that could extend the reach and impact of the campaign. Possible components could include press relations (including local, national and targeted pitching, Op-Eds, letters to the editor, editorial board visits, etc.), special events, grassroots marketing, customized web packages, mat release, multimedia newswire distribution, broadcast outreach, localized bites and B-roll feeds, and satellite and radio media tour.

The Ad Council unique approach to PSA campaigns creates campaigns exclusively through the creative services of a "volunteer" advertising agency that donates its labor pro bono and through donated media. However, all Ad Council campaigns incur certain hard costs. The campaign, for a 3-year effort, could cost $2.4 to $3 million. Specifically, these funds would pay for the hard costs associated with the campaign, including conducting qualitative and quantitative research; television, radio and print production; distribution (TV, radio, newspaper, magazine, Internet, out-of-home); media monitoring; research tracking studies; website development; public relations; and media outreach.

The funding for the 3-year effort would pay for two rounds of PSAs. Approximately 18 months after the campaign has been distributed to the media and thoroughly evaluated, the Ad Council and partnering organization would create and distribute new PSAs to the media. This approach will bolster the campaign's efficacy and maximize donated media since the media will increase their support when they receive newly created work.

Since the Ad Council relies on the pro bono services of ad agencies and media companies, the organization is able to produce national PSA campaigns for a fraction of the cost of paid media campaigns. Over 40 advertising agencies donate the creative work for Ad Council campaigns, and over 33,000 media companies contribute free ad space and time.

End Notes

[1] BLS Economic News Release: Employment Situation of Veteranc-2011, March 20, 2012, http://www.bls.gov/news.release/vet.nr0.htm

[2] http://www.themanufacturinginstitute.org/Research/Skills-Gap-in-Manufacturing/2011-Skills-Gap-Report/2011-Skills-GapReport.aspx.

[3] http://www.themanufacturinginstitute.org/Education-Workforce/Military-and-Veterans/Military-and-Veterans. aspx.

[4] The term "professional" implies an emphasis on non-research degree programs as a priority but not at the exclusion of strengthened advanced engineering research.

[5] There is increasing evidence of this need in many industries.

[6] The Ad Council is a private, non-profit organization that marshals volunteer talent from the advertising and communications industries, the facilities of the media, and the resources of the business and non-profit communities to deliver critical messages to the American public. The council produces, distributes, and promotes public service campaigns in support of key issue areas.

[7] http://www.abet.org/index.aspx.

[8] http://www.careeronestop.org/competencymodel/.

[9] BLS Economic News Release: Employment Situation of Veteranc-2011, March 20, 2012, http://www.bls.gov/news.release/vet.nr0.htm

[10] http://www.themanufacturinginstitute.org/Research/Skills-Gap-in-Manufacturing/2011-Skills-Gap-Report/2011-Skills-GapReport.aspx.

[11] http://www.themanufacturinginstitute.org/Education-Workforce/Military-and-Veterans/Milita ry-and-Veterans. aspx.

[12] The term "professional" implies an emphasis on non-research degree programs as a priority, but not at the exclusion of strengthened advanced engineering research.

[13] There is increasing evidence of this need in many industries.

[14] http://www.abet.org/index.aspx.

[15] http://www.careeronestop.org/competencymodel/.

In: Advanced Manufacturing in America
Editors: Brandon Cullerne and Michael Laston
ISBN: 978-1-62257-932-7

Chapter 6

CAPTURING A DOMESTIC COMPETITIVE ADVANTAGE IN ADVANCED MANUFACTURING. ANNEX 4: POLICY WORKSTREAM REPORT*

President's Council of Advisors on Science and Technology

EXECUTIVE SUMMARY

The United States is at risk of losing leadership in manufacturing. This is true not only in low-technology industries and products, but also with respect to our ability to manufacture the high-technology products that are invented and innovated in this country. We do not believe that it is the role of government to formulate a national industrial policy of direct investment in or subsidies to specific firms. However, we believe strongly that the United States requires a coherent national policy framework oriented toward creating a favorable business climate for manufacturing that spurs investments and fosters partnerships between government, academia, and industry. The manufacturing sector must have a competitive domestic environment, which includes a robust talent pipeline, physical capital, and intellectual capital.

While these capabilities are important to the overall health of the U.S. economy, they are particularly necessary for the advanced manufacturing sector, which faces intense global competition. Every major economic competitor is taking steps to attract investment in advanced manufacturing, recognizing the impacts of that investment on the broader economy.

Given the strong link between innovation and advanced manufacturing, the Partnership has put forward a package of synergistic recommendations to encourage greater U.S.-based innovation, research and development (R&D), and investment in advanced manufacturing. For the United States to continue to be an attractive location for businesses, it is important to build a policy framework that spurs investments and fosters partnerships between government, academia, and industry.

* This is an edited, reformatted and augmented version of President's Council of Advisors on Science and Technology, dated July 2012.

The foundation of that framework is constructed through targeted policies in four areas that have a significant impact on advanced manufacturing:

- Tax policy,
- Smarter regulations,
- Trade policy, and
- Energy policy.

The Advanced Manufacturing Partnership (AMP) Steering Committee (SC) has developed additional specific recommendations in three areas to improve the climate for robust industry/ university collaboration in research and commercialization, with the aim of reinvigorating what has been an historic strength of the United States:

- Remove barriers in tax-exempt buildings at universities to enable expanded university-industry collaborations,
- Expand capital resources for emerging advanced manufacturing enterprises, and
- Coordinate and expand existing programs that enable university researchers to interact directly with manufacturers.

Charge to the Workstream

The focus of the Policy Workstream is to create a policy environment that supports the overall objective of encouraging U.S.-based innovation, advanced manufacturing (and jobs), and international competitiveness and facilitating, where possible, government, academia, and industry collaboration. The workstream is also responsible for ensuring that the policy environment supports the work products of the other workstreams.

Process Followed

August/September: Workstream participants drafted a summary paper with objectives, questions, and a timeline as a starting point for reactions and contributions from AMP Steering Committee. This draft was followed by several conference calls. A survey of major U.S. academic researchers engaged in manufacturing policy was also completed.

October/November: The workstream gathered more feedback on policy issues and objectives at the regional AMP Steering Committee meeting at Georgia Tech. Additional specific input was solicited from workstream members. On October 24, 2011, a face-to-face meeting with about 25 participants was held to review everyone's input and to solicit additional suggestions and focus. A final series of conference calls was held with workstream members the week of November 14 to discuss and review the draft interim report.

December/January: The workstream developed specific near-term actionable recommendations to encourage U.S.-based innovations and to improve the climate for university/industry collaboration. A draft document of recommendations was generated for review by the full AMP Steering Committee.

February: The workstream refined the near-term actionable recommendations and finalized this document so it could be included with the final AMP Steering Committee report.

Key Findings

The United States is at risk of losing leadership in manufacturing. This is true not only in low-technology industries and products, but also with respect to our ability to manufacture the high-technology products that are invented and innovated in this country. We do not believe that it is the role of government to formulate a national industrial policy of direct investment in or subsidies to specific firms. However, we believe strongly that the United States requires a coherent national policy framework oriented toward creating a manufacturing climate that spurs investments and fosters partnerships between government, academia, and industry. To attract investment and production, the United States must promote a competitive business environment, which includes a robust talent pipeline, capital, a 21st century infrastructure, and strong investment in R&D. While these qualities are important to the overall health of the U.S. economy, they are particularly necessary for the advanced manufacturing sector, which faces intense global competition. Every major economic competitor is taking steps to attract investment in advanced manufacturing, recognizing the impacts of that investment on the broader economy.

Recommendations

Given the strong link between innovation and advanced manufacturing, the AMP Steering Committee has put forward a package of synergistic recommendations to encourage greater U.S.-based innovation, R&D, and investment in advanced manufacturing. For the United States to continue to be an attractive location for businesses, it is important to build a policy framework that spurs investments and fosters partnerships between government, academia, and industry. The foundation of that framework is constructed through targeted policies in four areas that have a significant impact on advanced manufacturing:

- Tax policy,
- Regulatory policy,
- Trade policy, and
- Energy policy.

A final section of recommendations is specific to promoting partnerships between universities, industry, and Federal laboratories and to enhancing industry/university collaboration in advanced manufacturing.

Tax Reform

A key focus of the AMP Steering Committee is on the important linkage between U.S.-based innovation, R&D, and manufacturing. To encourage investment in the United States, we must reform our corporate tax system to create a more attractive environment for business to be able to compete globally. The United States has the highest statutory corporate tax rate— including Federal and State taxes—among the 34 members of the Organization for Economic Co-operation and Development (OECD). This tax rate is an impediment to businesses that seek to invest in our country and for U.S.-headquartered businesses. Comprehensive U.S. tax reform is particularly important for the advanced manufacturing sector. Manufacturing is a source of direct and indirect high-paying jobs and is the underpinning of the U.S. middle class. Our current tax system discourages domestic capital investment in manufacturing, thereby undercutting the stability of the innovation and jobs engine that produced unparalleled economic prosperity during the last century. The tax system distorts investment by industry and asset type, with manufacturing, construction, and other high-wage industries paying a globally non-competitive statutory tax rate. The result is a decrease in aggregate investment in the sector. For these reasons, the tax system has to be reformed to address the existing distortions and disincentives for manufacturing in the United States. A more favorable tax climate would serve a two-fold benefit: incentivize increased investment by U.S.-based businesses and encourage more foreign direct investment in the United States, leading to an increase in investment, innovation, and jobs. Tax reform should also be internationally competitive with other tax systems in order to attract and retain advanced manufacturing and its associated innovation engine. While there is a need for broad tax reform to make U.S. companies more internationally competitive, our recommendations are more specifically targeted to the promotion of advanced manufacturing in the United States. They add up to an integrated package of proposals that address the mobile nature of capital and intellectual property (IP) and enhance the incentive for retaining and reinvigorating the historical strength of closely connected U.S. research and production capabilities. We believe that additional tax incentives should flow to those entities that engage in all three critical advanced manufacturing roles (U.S.-based innovation, R&D, and manufacturing):

- **Lowering the corporate tax rate to bring it more in line with other advanced economies.** A rate reduction, combined with broadening the tax base, would encourage additional investment in American manufacturing by U.S. corporations and would position the United States as a more attractive region for direct investment by foreign corporations.
- **Recognize the importance of manufacturing through the tax code.** Given global competition and the ripple effect of manufacturing on the economy, any tax reform should encourage investment in manufacturing. This can be achieved through a reduced tax rate for domestic manufacturing activity.
- **Strengthened and permanent R&D tax credits.** Increase the R&D alternative simplified credit to 20 percent and make it permanent.
- **Creating an internationally competitive corporate tax system.** Our tax system must be redesigned in a way that encourages companies to invest in the United States by addressing the current law on foreign earnings of U.S.-based companies. In

addition to lowering the overall corporate rate, reform must consider the tax treatment of overseas earnings of U.S.-based corporations, including consideration of a competitive partial exemption system similar to the type adopted recently by the United Kingdom or a minimum tax regime like the one in Japan. Ultimately, comprehensive tax reform must ensure that U.S. companies are competitive when operating abroad and in the United States.

The workstream recognizes that efforts to address long-term U.S. fiscal issues may bring about significant proposals that include a mix of rate reductions. The participants urge that this debate be particularly mindful of the imperative to recognize how this mix of actions may impact the climate for advanced manufacturing. The workstream also cautions against any new measures that could impede an improved climate for U.S.-based production or discourage investment in the United States.

Regulatory Policy

Regulation is an often criticized but vital function carried out by government. Well-conceived, science-based, and effectively implemented regulations are important tools for protecting consumers, workers, and the environment. When done right, regulation provides important societal benefits and can encourage greater competence and confidence in industry. Done excessively or inappropriately, or without unforeseen consequences in mind, regulatory policy can hamper innovation and international competitiveness. We recommend the following:

Early engagement. Collaboration between regulators and the impacted community can drive significant improvements in the quality of the final rules. Robust dialogue with agencies ideally should occur well before the comment period. Improved use of the advanced notice rulemaking process[1] would allow manufacturers to contribute to cost-benefit analyses in a meaningful way that could make compliance more cost effective.

Objective cost-benefit analyses. We recommend that cost-benefit analyses and risk assessments rely on the best available science.

Trade Policy

A fair and open international trading system provides the greatest opportunities for U.S.-based innovative manufacturing and, ultimately, for sustaining current jobs and creating new jobs. The U.S. Government needs to lead on a progressive trade policy, building on recent successes such as passing the U.S.-Colombia, U.S.-Korea, and U.S.-Panama Free Trade Agreements (FTAs). FTAs level the playing field for American exporters by eliminating tariff barriers to market access, reducing non-tariff barriers, and allowing access to dispute settlement systems. Trade policy is an important consideration for manufacturers choosing to build new facilities, but we must not let our competitors outpace us in the race to negotiate further agreements. The United States must prioritize policies that help ensure access to foreign markets and promote global competitiveness. These policies must include a focus on non-tariff barriers and export control policies. The Trans-Pacific Partnership (TPP) is an example of a high-standard, ground-breaking negotiation that will cover new emerging barriers for cutting-edge technologies, promote regulatory coherence, address competition with state-owned enterprises, and provide a template for economic integration across Asia Pacific. In balance with trade liberalization, the U.S. Government should focus strongly on enforcing trade rights, particularly those addressing market-distorting subsidies, unfair trade practices, and IP violations to level the playing field for U.S.-based manufacturing.

As a near-term goal, the U.S. should:

- **Pursue increased market access.** The future key barriers are not tariffs. They are non-tariff barriers—regulatory and standards impediments that represent de facto market barriers. Examples of non-tariff barrier areas are innovation principles, regulatory reform and customs facilitation, forced technology transfer, and IP enforcement/counterfeiting. The Federal Government must strengthen the interagency process to create a consistent agenda on regulatory issues. The U.S. Government should strengthen cooperative, capacity-building initiatives with other key trading partners.
- **Launch new negotiations.** The U.S. Government has actively solicited input from industry on core economic trading partners for new negotiations. A number of regions, such as the Middle East and North Africa, can benefit from near-term capacity-building efforts to lead to eventual full trade liberalization efforts. In the interim, the U.S. Government should prioritize a Trans-Atlantic Partnership (TAP) negotiation, which would leverage the advanced economies of the United States and the European Union (EU) and allow both to address 21st century trade barriers (such as regulation, innovation, etc.) as a model for future multilateral trade liberalization.
- **Reform export controls.** The U.S. Government is making progress in reforming outdated export control regimes—starting with rebuilding the U.S. Munitions List—by harmonizing the export control licensing and administration procedures across all involved agencies and transitioning all involved agencies to a single information technology (IT) platform. The U.S. Administration should accelerate this work and actively incorporate industry input in a modernized export controls regime.

Energy Policy

Energy is a basic building block for today's advanced manufacturing applications. Advanced manufacturing uses innovative technologies to add value to raw energy inputs in order to produce modern materials and solutions, including electronic materials, pharmaceutical breakthroughs, and clean energy alternatives. However, U.S. energy policy must fully account for the impacts of energy costs on manufacturers and the potential to drive investment into new markets and applications as the United States seeks to transition to a sustainable energy future. Therefore, any effort to reinvigorate advanced manufacturing in the United States would not be complete without an examination of energy policy that seeks ample supplies to catalyze economic growth and prosperity.

We recommend the following:

- **Focus on energy efficiency and conservation.** Energy efficiency is the most affordable and available way to lower energy costs and reduce carbon emissions and is particularly important to the manufacturing sector. Every dollar saved through energy efficiency efforts can be redeployed to expand business and preserve manufacturing jobs. For example, according to the Brookings Institute, if all eligible buildings in the United States were retrofitted over the next decade, roughly 215,000 direct jobs—127,000 of which are in manufacturing—would be created. We recommend policies that provide incentives for power generators and distributors to undertake cost-effective and innovative energy efficiency measures and promote tools to assist all manufacturers in implementing energy efficiency measures.
- **Increase and diversify domestic supplies.** American economic growth will continue to rely on hydrocarbon energy (e.g., oil, naphtha, natural gas, ethane, or coal) and will require additional domestic supplies to improve energy security and reduce price volatility. These inputs are critical for the manufacturing process as both fuel and feedstock, serving as the basic building blocks of materials used in 96 percent of all manufactured goods, including products enabling the further development of renewable sources of energy, such as solar panels and wind blades. Onshore, increased supply from unconventional sources, such as natural gas, oil, and natural gas liquids from shale, will be important resources for the United States over the next several decades. The availability of these resources for value-added products must be a policy imperative to ensure economic growth and job creation. Producers and regulators need to work together to ensure that potential reserves can be brought to market in an environmentally acceptable manner at an affordable cost. Natural gas at stable, competitive prices will continue to incentivize American manufacturers to invest and create jobs in the United States. Today, industrial uses of natural gas as a feedstock are driving multibillion-dollar investments. In turn, multiplier effects from these investments will be felt across the economy, including other U.S. manufacturers less dependent on hydrocarbon feedstocks.
- **Speed development of renewable sources of energy.** Government, industry, and academia have roles in accelerating the development of effective and more sustainable alternative energy sources, including renewable sources. As global demand for clean sources of energy grows, the United States has the opportunity to play a key role in the manufacturing of advanced technologies, such as energy storage, photovoltaics, and wind power. Since 2008, the United States has nearly doubled renewable energy generation. In 2011, U.S. solar installations grew 109 percent, with the overall solar market surpassing $8.4 billion. However, renewables remain a small fraction of U.S. energy use. Policies that primarily focus on driving down costs are needed. Lower costs, in turn, will help drive increased demand. We recommend the continued extension of financial incentives for public/private research into promising technologies and storage devices. Further, any incentives that spur the early adoption of innovative technologies, such as low- and no-carbon sources originating from coal, solar, natural gas, wind, tidal, and geothermal energy, must be targeted at technologies that demonstrate a path toward economically viability.

Transition to a low-carbon economy. To create a sustainable energy future over the long term, we believe that the United States needs to shift to a low-carbon economy. The right mix of fundamental research, innovation, and aggressive implementation is needed to achieve this transition and continued economic growth. The development and implementation of a broad portfolio of technologies are essential for this transition. The United States has the technical capacity to accelerate development of sustainable energy options, but large-scale commercialization of new capital-intensive manufacturing solutions will require increased public/private partnership. We recommend a targeted approach to promote aggressive basic research and development, with accelerated demonstration and deployment of clean energy and new-generation energy-efficient technologies. Government policy can help most in specific situations, such as when the costs and market development risks of critical technology exceed the commercial capabilities of individual companies, where the regulatory or liability risks are beyond the capacity of the private sector, and when investment timelines exceed the private sector's capabilities.

Empowering Enhanced Industry/University Collaboration in Advanced Manufacturing

One of the most significant hurdles to a robust manufacturing sector is the disconnect between U.S. manufacturers and the innovation dynamics that U.S. universities have developed following the Bayh-Dole Act. In order to overcome this, there is a need to address the fundamental barriers that impede small- and medium-sized manufacturing firms from engaging with university research and to ensure that an adequate technical talent base is developed to support innovation in advanced manufacturing. This approach includes exploiting the opportunities for synergy between initiatives designed to improve the overall climate for U.S. manufacturing and the environment for industry/university collaboration.

We have examined opportunities to accelerate the development of effective research and licensing agreements and have explored how agency research programs can be enhanced to contribute to this reinvigoration of industry/university collaboration.

Revenue Procedure 2007-47

The current Revenue Procedure 2007-47 restricts the "private business use activities undertaken in universities buildings financed with tax-exempt bonds. Under these provisions, particularly sections 6.02 and 6.03, industry-sponsored research is considered a private business use unless the university obtains a fair market value for the outcomes of the research. Specifically, these provisions state that fair market value must be determined at the time the license or resulting technology is available for use.

Exceptions to these private use restrictions include licenses awarded to consortia of companies and non-exclusive licenses that provide similar use rights to all users. Should private uses exceed the limit (10 percent for public universities and 5 percent for private universities—though when bond transaction costs are included, the real limits are 8 percent and 3 percent), the tax-free status of the bonds may be revoked. Private-use activities also covered under the total space cap would include bookstores and other retail activities, such as coffee shops and restaurants. It also includes space allocated by universities to early-stage spin-out companies to undertake early research activities in labs of buildings with tax-exempt bond financing and space allocated for visiting industrial researchers.

These provisions have clearly had the effect of incentivizing universities to adopt policies of industrial collaboration that preclude exclusive licenses or stipulate that commercial licensing terms must be negotiated at the conclusion of the research. More specifically, these provisions directly limit the ability of universities to enter into research agreements with companies that include upfront provisions for exclusive licenses to the resulting technologies.

These tax provisions and restrictions are unique to U.S. universities and create a competitive disadvantage at a time when research is increasingly a global enterprise and when a reinvigoration of industry/university interaction could contribute to securing leadership and domestic production in advanced manufacturing technologies and contribute to the robust in-sourcing of manufacturing. The restrictions directly conflict with the following imperatives that shape competitiveness in advanced manufacturing:

- 2007-47 adds friction and uncertainty to industry/university collaborations at a time when hyper competition rewards speed and flexibility. The inability to have certainty over commercial rights negotiations until after the completion of the research is a clear disincentive for U.S. manufacturers to pursue more strategic relationships with universities.
- 2007-47 reflects a linear view of the research continuum at a time when discovery often involves more iterative processes of interaction between fundamental and developmental advances. In essence, 2007-47 envisions a clear departure point between basic and applied research while emerging applications in highly interdisciplinary areas often involve a more integrated mix of basic and applied development.
- 2007-47 unintentionally creates the potential for competition between university commitments to support and nurture spinouts and industrial research agreements. Strategies for incubating early stage startups in university labs, for example, compete directly with industry-sponsored research agreements that have upfront licensing terms for space under the cap limit. This competition is clearly not in the public interest and comes at a time when the greater imperative is to foster synergy between the university start-up engine and the scale-up strengths of U.S. manufacturers.
- 2007-47 creates a barrier in allowing visiting industrial and Federal Government researchers to be embedded on university campuses to promote active collaboration.
- The Internal Revenue Service (IRS) recently completed a detailed audit on the cap allocation practices of 30 universities to guide further rulemaking—suggesting that the trend may be toward greater restrictions.

From the Morrill Act to the rapid development of collaborations by the Office of Scientific Research and Development (OSRD), a long history of policy initiatives have enabled American industry and universities to collaborate effectively in the face of changing circumstances. The proposal to create a waiver from or revise 2007-47, while not as grand in scale as the examples cited previously, is nonetheless designed to make a tangible impact on the day-to-day workings of research collaboration and create the context for more expedient partnership development capable of accelerating commercialization. The proposal is designed to remove policy barriers and, more specifically, the policy "fog" that currently surrounds the development of collaborations, creating a context for more transparent strategies and speedier development. A waiver from 2007-47 will not create uniform IP policies across all U.S. universities. The diversity of approaches—reflecting the variety of missions and focus that has been an historic strength—will continue to exist. A waiver from 2007-47 does not create a new bureaucratic process. It leverages an existing one. The provisions for Form 990 submissions requiring the existence of a space allocation plan and monitoring capacity establish a compliance burden on universities that the waiver uses to reduce friction in the partnership process. The proposal seeks to capture a full and robust vision of collaboration that extends from sponsored research to greater synergy between industry engagements and university start-up activities. Finally, the proposal levels the international industry/university research playing field.

- Establish a waiver from Revenue Procedure 2007-47 to enable expanded university/industry collaborations. Enable universities to apply for a waiver from Revenue Procedure 2007-47 restrictions on private use activities in buildings constructed with tax-exempt bonds for the specific purposes only of expanding industrial R&D collaborations. Universities would be eligible to apply for a waiver from the limits on private business use restrictions outlined in provision 2007-47 for specific industrial R&D collaborations only. This application would be developed under existing requirements for each university to have a formal plan and management strategy for monitoring the allocation of all private business use space under the limits that are mandated as part of the annual preparation of Form 990 and would be submitted with a copy to the Advanced Manufacturing National Program Office (NPO). Implementation should include a White-House-facilitated review by the Department of Treasury and the IRS Office of the Associate Chief Counsel. A model waiver request should be developed and should include the following:
  - An outline of specific strategies for industry partnerships for R&D. These strategies should include agreements containing exclusive rights provisions negotiated as part of the sponsored research and/or partnerships with industry to accelerate new business development that includes but is not limited to business-sponsored incubators for early stage startups, including university spinouts and joint venture companies and/or space allocations for fostering visiting industry and government scientist programs and/or the allocation of university labs and/or specialized equipment for access by small- and medium-sized industrial firms.
  - A clearly outlined strategy for developing and implementing the partnerships, including the development of standard licensing agreements for industry-sponsored research and faculty spinouts and demonstrated provisions for the ready adoption of industry master agreements.
  - A clearly outlined plan for engagement with industry in the development of partnerships to be covered under the waiver plan.
  - Clear evidence that the waiver is not designed to simply augment expansion of non-industry development partnerships and activities in tax-exempt bond space.

Capital Resources for Emerging Advanced Manufacturing Enterprises

A wide-range of issues relate to U.S. manufacturing's access to capital; however, in keeping with the AMP Steering Committee's focus on capturing domestic competitive advantage in newly emerging advanced manufacturing technologies, this outline seeks to highlight potential actions in three specific points on the development continuum. The first point in the funding continuum that these recommendations seek to address is the gap between the conclusion of basic research and early stage funding—the pre-seed funding necessary to develop prototypes and early stage market exploration and validation tools. The second point in the funding continuum that these recommendations seek to address is the funds needed for early stage scaling up of production to support activities (e.g., beta unit development) that are often required before large-scale financing can be secured from traditional sources. One critical opportunity in this regard is to identify a means of fostering greater synergy among university startups and larger manufacturing firms to take advantage of the scale-up capabilities that these firms bring to bear. The third point in the funding continuum that these recommendations seek to address are some concepts put forward to expand available funding to move from early stage scale-up to pilot plant development. The specific recommendations that follow are made for consideration.

Increase the Pipeline of Start-Ups in Advanced Manufacturing

Building upon a proposal developed by the nation's major university presidents for the National Advisory Council on Innovation and Entrepreneurship (NACIE), a recommendation is made to create a Phase 0 Small Business Innovation Research (SBIR) program for major research areas in advanced manufacturing. This program would provide support for the critical pre-early stage funding activities associated with testing the commercial potential of new technologies, including early prototype development and market development.

States such as Florida and Nebraska have worked with their universities to develop Phase 0 programs. Often these programs focus on helping already established start-ups prepare for formal SBIR applications. In addition to this focus, an advanced manufacturing Phase 0 program would focus on helping companies at the formation stage develop beta test results and customer relationships. In addition, greater attention to comply with Executive Order (EO 13329) issued by President Bush in 2004, requiring SBIR and Small Business Technology Transfer (STTR) programs to give high priority to manufacturing-related R&D projects would contribute to a more vibrant base of support for manufacturing related start-ups.

Expand the Resources Available for Early Stage Growth and Accelerate Start up Interaction with Major Manufacturers

Some models designed to build upon innovative public/private partnerships have created stronger support to enhance the growth of early stage companies emerging from Federal research funding. For example, the National Science Foundation (NSF) created a 501(c)3 not-for-profit corporation (Innovation Accelerator) to expand resources to its SBIR award winners. Thus, far Innovation Accelerator has helped to establish a regional seed fund for NSF SBIR firms and has provided mentoring assistance and aid in leveraging $200 million in investment for these firms. The recommendation is to expand the Innovation Accelerator program to support start-ups emerging from Federal advanced manufacturing research programs. Support from Innovation Accelerator could include expanded mentoring initiatives, efforts to mobilize corporate partners interested in supporting beta testing of products, and/or the creation of a seed fund for advanced manufacturing start-ups.

By far, the strongest resource to nurture the growth of early stage advanced manufacturing companies is to increase the synergy between these firms and established U.S. manufacturers. This kind of start-up/established firm interaction is common among leading IT companies and in the life sciences but is less common in traditional manufacturing areas. The core of these interactions often involves activities that are not directly related to financing and are frequently vital pre-conditions to early stage and pilot plant development. These activities include concept validation that draws upon the broad market and production operations of leading manufacturers, use of specialized facilities, business and production mentoring, and early beta customer partnerships. These interactions can also lead to investment relationships from venture operations of major companies. In the ongoing evaluation of policies for advanced manufacturing, future policy reviews should also explore the tax treatment of investments by leading manufacturing firms in start-ups in the context of overall tax competitiveness improvements.

A focus on creating a national network to foster stronger start-up interactions with leading manufacturers would be a key element of the expanded mission of Innovation Accelerator. This network would complement and aid the efforts of individual universities to integrate a manufacturing focus into innovation ecosystem programs to create and support start-ups.

Clear the Pathway from Start up to Pilot-Scale Production: Align to Full-Scale Market Development

One area often cited by technology based start-ups as a source of competitive disadvantage is the direct funding that many international competitors make available for pilot plant and early stage tool development and other fundamental building blocks needed to secure the financing for full-scale production. While the United States does not have a direct counterpart to these programs, some very successful models of strategic procurement initiatives directly support the growth of production capabilities. The most effective U.S. counterpart to the "ramp-up funding international competitors provide may be the Defense Production Act Title III funding. The Title III program is intended to provide the Department of Defense (DOD) "a powerful tool to ensure the timely creation and availability of domestic production capabilities for technologies that have the potential for wide-ranging impact on the operational capabilities and technological superiority of U.S. defense systems. In recent years, investments under Title III have included projects in automated composites production, thermal battery industrial infrastructure, and biofuels production. Examining opportunities for advanced production procurement set-asides in other areas of Federal operations could yield a similar roster of diverse manufacturing-related applications that could benefit from a competitive program that supports early market demand and helps establish a production supply chain. The key to the effectiveness of this approach is the ability to foster close interaction between acquisition and procurement operations and research initiatives in the early stages of technology development. A recommendation is made to create a formal collaboration between the Advanced Manufacturing NPO and DOD Title III and other relevant Federal procurement programs to identify opportunities for targeted procurement set-aside competitions that can create pathways to market growth beyond early stage funding and to help ensure that advanced manufacturing breakthrough technologies are integrated into a U.S. supply chain.

Keep Our Start-Ups Making It at Home: Strategies for Building Manufacturing Support Activities into University Innovation Ecosystems

The last 30 years has witnessed a dramatic transformation in university technology transfer operations and specifically in the emergence of comprehensive university infrastructures or ecosystems to encourage, support, and accelerate the development of spin-out companies. The emergence of these start-up ecosystems is a vital component of the competitive advantage that U.S. research universities provide to the American economy.

In its summary of technology transfer activity for Fiscal Year (FY) 2010, which was released in July 2011, the Association of University Technology Managers (AUTM) reported the following key indicators of this vitality for the 183 U.S. universities that reported results:

- 651 startup companies formed, 498 of which had their primary place of business in the licensing institution's home state—a more than 10 percent increase in start-up activity over the previous year,
- 657 new commercial products created, and
- 3,657 startups still operating as of the end of FY 2010.

The core components of this start-up engine are an extensive web of programming and support functions that universities have developed to increase the rate of new business formation by faculty and students. Among the activities that make up university innovation ecosystems are student and faculty entrepreneurship training programs, mentoring programs, entrepreneur-in-residence programs to link faculty to business expertise, pre-seed and seed funds, venture capital attraction, and incubator and often extensive partnerships with regional economic development organizations to provide additional resources for early stage growth.

The emergence of these comprehensive innovation ecosystems over the last 30 years has left U.S. universities without rivals in their ability to transform research into new business. However, these ecosystems have tended to be less structured in creating pipelines to domestic manufacturing opportunities and in mobilizing the resources and networks needed to connect start-ups to U.S. production.

Building a strong connection between university start-ups and the U.S. manufacturers and suppliers can make a vital contribution to ensuring future leadership in advanced manufacturing. Moreover, the development of a manufacturing focus in university innovation ecosystems offers additional opportunities for strengthening partnerships with U.S. manufacturers that, as noted previously, have vital resources to contribute in terms of capabilities to help scale up production.

The following actions could make a vital contribution to building a manufacturing focus into the university start-up support environment:

- **Keep score.** Incorporate manufacturing-related outcomes into university technology transfer benchmarks. A critical first step to reinforce the importance of building stronger linkages to manufacturing resources is to begin measuring and highlighting university success in capturing domestic production from technology transfer activities. Incorporating into annual AUTM reports a manufacturing impact component that reflects domestic production captured from start-ups and licensing activity would place manufacturing front and center in university technology transfer strategy development. Including these measures in the annual AUTM survey would also help stimulate a vibrant exchange on best practices and would encourage greater focus on manufacturing in regional economic development partnerships and among universities and manufacturers in the development of sponsored research partnerships.
- **Build stronger linkages between manufacturing support resources and university innovation ecosystems.** In addition to developing new measures to capture the success of universities in realizing the domestic production potential emerging from start-ups, a clear opportunity exists to identify some very specific activities to integrate a manufacturing focus into university innovation ecosystems. A first step in advancing these connections is to expand the work of the nation's Manufacturing Extension Partnerships (MEPs) to create direct supply-chain development, prototyping, and early stage engineering services for university start-ups. Several pilot programs have been launched by individual MEPs and work has begun on creating a national network to support the identification of potential suppliers for start-ups. These initial steps should be expanded, with the objective of creating the types of embedded connections to university technology transfer operations that early stage investment and incubation programs have fostered.

  One potential area of focus of these expanded MEP connections would be to help ensure that three-dimensional (3D) production design and other next-generation "make" technologies are standard resources for early stage spin-out entrepreneurs. The specific objective is to ensure that every MEP has designed and implemented a formal start-up support strategy in collaboration with universities in their region. These strategies would formalize specific approaches to indicate how the MEPs and universities would work together to bring a manufacturing business plan development to the earliest phase of start-up creation.

  Other partners can augment the role of MEPs. Organized labor brings unique resources for helping identify domestic production opportunities, and, as highlighted several times in this discussion, the scale-up strengths of leading manufacturers represent a unique resource for opening pathways to domestic production for university start-ups.

Recognizing U.S. manufacturers as start-up support resources and as research partners would help reshape innovation ecosystems and add new dimensions to the historic relationship between industry and universities.

- **Implementation.** To begin speedy implementation of this recommendation, universities should direct their technology transfer managers to work with the AUTM's leadership to help develop appropriate and effective measures of impact on domestic manufacturing. In developing these measures, the universities' AUTM representatives can also craft a strategy to launch measures that can complement the goal of elevating attention to manufacturing, which pervades a number of the AMP Steering Committee recommendations.

  To begin ensuring that specific manufacturing development services are standard components of all university innovation ecosystems, universities can work with the Advanced Manufacturing NPO to convene a meeting of MEPs and university technology transfer representatives. The purpose of this meeting would be to ensure broad-based awareness of existing pilot programs and supplier network development. The session could also lead to the development of a template for MEP/university collaboration strategies and feature best practice exchanges on strategies for engaging manufacturers and labor organizations as partners to support domestic production development.

Establish a "Make for America" Initiative

Create a coordinated inter-agency initiative under the direction of the Advanced Manufacturing NPO to expand existing programs that enable students, faculty, and post-doctoral researchers to interact directly with manufacturers and to establish a 501(c)3 organization to foster public/ private support for this expansion.

Both the Policy and the Education and Workforce Development Workstreams identified in their interim reports and recommendations the critical need to increase engagement between students, faculty, and post-graduate researchers and manufacturers, which this initiative seeks to integrate. Expanding this engagement provides underlying support to a host of strategic objectives. More robust interaction in the form of visiting programs can contribute to changing the image of manufacturing, to encouraging a broader recognition of the national imperative for U.S. leadership in advanced manufacturing, to increasing interest in formal curricula and degree programs, and significantly and rapidly to augmenting the ability of research universities to contribute to small manufacturing firms. Moreover, since major research engagements often begin with a visit or a conversation, expanding faculty and student interaction with companies can be a vital stimulant to increasing the pipeline of collaborations that other policy recommendations are designed to enhance.

In addition, a coordinated and enhanced initiative can directly complement related proposals—such as the Jobs Council recommendation for "community grand challenges" to improve the local environment for manufacturing firms. More pointedly, it is simply time to involve manufacturers in the kind of dynamic, often informal engagement that is the hallmark of interactions between universities and Internet and IT firms.

Effective Federal Government efforts are in place to build upon programs already in existence—most notably GOALI, the NSF-directed Grant Opportunities for Academic

Liaison with Industry. However, these efforts are spread across agencies with little coordination, which restricts the overall ability to convey real engagement and build excitement regarding manufacturing as a focus for research, education, and a career. We propose to create an interagency umbrella for energizing these efforts and to connect them seamlessly to private-sectorled initiatives, such as the Jobs Council initiative to spur the education of 100,000 engineers. Building upon models used in other policy initiatives, such as the efforts to better attract and support NSF SBIR recipients and to nurture the growth of education technology start-ups, this proposal calls for linking a clear inter-agency leadership directive with the formation of a not-for-profit corporation that can nimbly collaborate with private sector, foundation, and state and local government partners.

The Advanced Manufacturing NPO will be designated as the lead for providing inter-agency coordination of programs to facilitate university student, faculty, and post-doctoral engagement with manufacturing firms. In this role, the NPO will

- Establish a clear and compelling national numeric goal for a mobilizing and obtaining expanded internships and fellowships with companies that focus on advanced manufacturing;
- Develop a definition of the elements that would constitute a "Make for America" internship in order to allow enable existing programs to become affiliated and engaged;
- Develop alternative models for manufacturing internships and fellowships;
- Identify opportunities for synergy among agency program operations;
- Coordinate and promote internships and fellowships associated with manufacturing innovation institutes and create an umbrella Make for America marketing and recruitment strategy that elevates awareness and excitement for such programs;
- Make recommendations on long-term strategies for developing broad and strategic internship and fellowship programs that engage a diverse set of manufacturing sectors, regions of the nation, types of universities, majors of students and expertise of faculty; and
- Identify opportunities for enhanced internship and fellowship initiatives to strategically complement national technology development, commercialization, and small manufacturer competitiveness objectives. For example, create regional manufacturing "boot camps" for interns and fellows to provide an enhanced educational and networking experience.

Under the direction of the Advanced Manufacturing NPO, key steps include completing a detailed assessment of existing programs, exploring the potential for inter-agency coordination, engaging private sector and university partners building off of the AMP, investigating the value of an umbrella branding effort for elevating the programs, expanding interest and attracting additional resources, and examining the direct value of creating a not for profit corporation.

SUPPORTING MATERIALS

Literature Review

Since the spring of 2011, a number of excellent high-level reports have been prepared through a collaborative process involving government, industry, academia, and others whose recommendations are relevant and highly useful here. The following is a brief summary of noteworthy comments and highlights that directly relate to the nexus of U.S.-based innovation, R&D, and manufacturing.

- **President's Jobs Council.** While claiming that there is no silver bullet," the council focused on regulatory reform and pushed for job-rich projects in infrastructure and energy, the promotion of entrepreneurial high-growth (jobs) firms, investments designed to increase U.S. jobs, a simplified regulatory review, and talent development. Their report also touched on increasing travel to the United States and streamlining inward investment and investor immigration. Improving the medical device approval process was also highlighted.
- **President's Council of Advisors on Science and Technology (PCAST).** This council focused on taxes and innovation. The council called for investment in shared infrastructure facilities with universities, development of advanced manufacturing processes, and partnerships with industry/academia in broadly applicable future technologies, such as nano-manufacturing, IT, and advanced materials. The three compelling reasons cited for this investment in the PCAST support were (1) jobs, (2) innovation, and (3) security.
- **Commerce Secretary's Manufacturing Council.** This council's top recommendations were in the areas of tax reform, regulatory reform, and innovation and R&D. Specifically, they sought a 25 percent or lower tax rate, a territorial tax, and a permanent R&D tax credit. They also recommended external regulatory collaboration, benchmarking and streamlined regulations, and compliance. Regarding innovation, they pushed for increased investment in basic and applied technologies.
- **President's Export Council (PEC).** This council has proposed a number of detailed letters of recommendation focusing on trade (FTAs, Russia's accession to the World Trade Organization (WTO), and the TPP/21st century trade model) and on infrastructure development (investment prioritized by export value), work-force readiness, visa reform, and progress on export control reform. The PEC has gone further to develop a measurement device (stop/light chart) to track progress on the PEC's recommendations.

End Note

[1] Advanced rulemaking is intended to solicit comments and information from all segments of the public interested in a particular issue before an agency determines whether a rule (regulation) will be proposed.

In: Advanced Manufacturing in America
Editors: Brandon Cullerne and Michael Laston ISBN: 978-1-62257-932-7

Chapter 7

CAPTURING A DOMESTIC COMPETITIVE ADVANTAGE IN ADVANCED MANUFACTURING. *ANNEX 5: OUTREACH WORKSTREAM REPORT*[*]

President's Council of Advisors on Science and Technology

ABSTRACT

In June 2011, the President established the Advanced Manufacturing Partnership (AMP), which is led by a Steering Committee that operates within the framework of the President's Council of Advisors on Science and Technology. In July 2012, the AMP Steering Committee delivered its report to PCAST, entitled *Capturing Domestic Competitive Advantage in Advanced Manufacturing*. PCAST adopted this report and submitted it to the President. The Steering Committee's report draws on preliminary reports prepared by several "workstreams." These workstream reports have been made available as on-line annexes to the Steering Committee report.

EXECUTIVE SUMMARY

Between September 2011 and March 2012, the Outreach Workstream held a series of meetings with stakeholders who have an interest in advanced manufacturing policy. Stakeholders included Congressional staff, trade associations, scientific associations, and think tanks. These meetings had three purposes: (1) to educate stakeholders about the Advanced Manufacturing Partnership (AMP), (2) to solicit feedback about issues on which the AMP Steering Committee (SC) should focus, and (3) to learn about what their respective organizations are doing that could inform AMP's work and be a resource for the Advanced Manufacturin gNational Program Office (NPO).

---

[*] This is an edited, reformatted and augmented version of President's Council of Advisors on Science and Technology, dated July 2012.

KEY FINDINGS

- Strengthening manufacturing has bipartisan support in Washington, DC. Although people quibble about the details and the exact role of the Federal Government in strengthening the Nation's manufacturing sector, there is bipartisan and broad agreement that an ongoing public/private partnership to enhance manufacturing is a worthwhile effort.
- The Nation should prioritize Federal research and development (R&D) investments to ensure that they are closely tied to manufacturing and are "intellectual property (IP) dense." If something is easily replicable, it can be done other places.
- There is no silver bullet to growing and keeping advanced manufacturing in the United States. The Nation needs a mixed approach for success—workforce skills, roadmaps, R&D, all collaborative.
- The Nation should take better advantage of the Manufacturing Extension Partnership (MEP) program and other Federal programs that are well positioned to help existing manufacturers innovate. This is especially important for small- and medium-sized manufacturers. The Advanced Manufacturing NPO should figure out a way to take the MEP to the next level.
- The Nation must do a better job of crafting Federal workforce programs to support advanced manufacturing. There is some skepticism about whether Workforce Investment Act (WIA) and other Department of Labor (DOL) programs are doing much to help workers become qualified for jobs in the manufacturing sector.
- Manufacturing has an image problem in the United States. The Advanced Manufacturing NPO should bring together interested stakeholders to work on changing the cultural attitudes about manufacturing.

RECOMMENDATIONS

- **Develop a long-term Image of manufacturing awareness campaign.** Engage the Ad Council to develop a Madison Avenue approach to improve the image of manufacturing businesses and employees. The Advanced Manufacturing NPO should manage the initiative.
- **The Advanced Manufacturing NPO needs to be a coordinator of Federal activities related to advanced manufacturing.** As disparate agencies advance their own investments related to advanced manufacturing, the NPO needs to serve as a coordinator of all Federal activities and the authority for external audiences about the Federal Government's advanced manufacturing priorities.
- **Develop an outreach program.** Use national engineering and professions organizations and regional chapters of organizations such as the Institute of Industrial Engineers (IIE), Society of Manufacturing Engineers (SME), and the National Association of Manufacturing (NAM) to work directly with high schools, community and technical colleges, universities, and states to spread the messages of the importance of manufacturing.

In: Advanced Manufacturing in America
Editors: Brandon Cullerne and Michael Laston
ISBN: 978-1-62257-932-7

Chapter 8

CAPTURING A DOMESTIC COMPETITIVE ADVANTAGE IN ADVANCED MANUFACTURING. *ANNEX 6: REGIONAL MEETING SUMMARIES*[*]

President's Council of Advisors on Science and Technology

ABSTRACT

In June 2011, the President established the Advanced Manufacturing Partnership (AMP), which is led by a Steering Committee that operates within the framework of the President's Council of Advisors on Science and Technology. In July 2012, the AMP Steering Committee delivered its report to PCAST, entitled *Capturing Domestic Competitive Advantage in Advanced Manufacturing*. PCAST adopted this report and submitted it to the President. The Steering Committee's report draws on preliminary reports prepared by several "workstreams." These workstream reports have been made available as on-line annexes to the Steering Committee report.

SUMMARY OF ADVANCED MANUFACTURING PARTNERSHIP REGIONAL MEETINGS PRESIDENT'S COUNCIL OF ADVISORS ON SCIENCE AND TECHNOLOGY ADVANCED MANUFACTURING PARTNERSHIP STEERING COMMITTEE

Meetings Held: October - December, 2011
Report Date: February 1, 2012

[*] This is an edited, reformatted and augmented version of President's Council of Advisors on Science and Technology, dated July, 2012.

1. Overview

The Advanced Manufacturing Partnership (AMP) is a national effort bringing together the Federal government, industry, universities, and other stakeholders to identify and invest in emerging technologies with the potential to create high-quality domestic manufacturing jobs and enhance the global competitiveness of the United States. AMP responds to recommendations made by the President's Council of Advisors on Science and Technology (PCAST) in its June 2011 report, entitled "Ensuring Leadership in Advanced Manufacturing." AMP is guided by a Steering Committee, which operates within the framework of PCAST and is comprised of leading experts from industry and academia. The AMP Steering Committee (AMP SC) is organized by four workstreams: Technology Development; Policy; Education and Workforce Development; and Shared Facilities and Infrastructure. In the fall of 2011, the AMP SC hosted four regional outreach meetings around the country and over 1000 members of the public representing diverse stakeholder perspectives participated:

- Atlanta, GA, October 14, 2011 (hosted by the Georgia Institute of Technology)
- Cambridge, MA, November 28, 2011 (hosted by Massachusetts Institute of Technology)
- Berkeley, CA, December 5, 2011 (hosted by University of California - Berkeley and Stanford University)
- Ann Arbor, MI, December 12, 2011 (hosted by the University of Michigan)

The regional meetings had two purposes. The first was to share AMP's approach and activities with industry, university, government, and other stakeholders. The second was to gather the participants' ideas about opportunities for investments and actions that have the potential to transform manufacturing in the United States. Participants at each of the four regional meetings heard from government and industry panelists, who discussed opportunities and challenges that their communities face in the advanced manufacturing domain. These panel discussions were followed by breakout sessions focused on the themes of the four workstreams. The main points of these breakout sessions are synthesized below. Specific details on each meeting can be found in the Appendix.

AMP Workstreams

Technology Development identifies emerging technologies with transformative potential with the express intent that they be commercialized and deployed in the United States.

Policy makes recommendations to the Administration on economic and innovation policies that can directly or significantly impact the ability to improve research collaboration and the pathway to commercialization in support of U.S. based manufacturing and jobs.

Education and Workforce Development identifies tangible actions that will support a robust supply of talented individuals to provide human capital to companies interested in investing in advanced manufacturing activities in the United States.

Shared Facilities and Infrastructure assesses opportunities to de-risk, speed up, and lower the cost of accelerating technology from research to production through unique capabilities and facilities that serve all U.S. based manufacturers, in particular small- and medium-sized manufacturers.

A. Technology Development Workstream

Each regional meeting approached technology development differently. At the first regional meeting in Atlanta, the participants in the technology development group discussed which advanced manufacturing technologies were critical to develop to enhance U.S. competitiveness in advanced manufacturing and innovation. Participants cast a wide net and identified areas of urgent importance, including energy (old and new sources, efficiency, and innovative technology); microelectronics; informatics (especially in health care); materials (composites, nanomaterials, biomaterials) and technology for forming and shaping materials; supply of strategic materials (including rare earth materials); green technology; and broadly applicable systems (automation, supply chain models, predictive modeling, complex systems, and information management). At subsequent meetings, the discussions were launched from a narrowed list of topics that cut across multiple industries and technology domains. These topics reflect the evolving discussions of the technology development workstream. Over the course of the four regional meetings, the ideas discussed by the participants evolved and were refined resulting in five possible cross-cutting technology domains:

- Advancing sensing, measurement, and process control
- Advanced Materials Design (including nanomaterials, metals, coatings and ceramics)
- Information technologies (including visualization)
- Energy efficient manufacturing
- Nanomanufacturing

B. Policy Workstream

The main themes discussed in the policy workstream breakout sessions included regulatory burdens, intellectual property, export controls, trade policy, and tax policy. While many companies recognize the regulations are important and needed, they contend that the process is often burdensome and lacks coordination between different Federal and state requirements. Policies that assist manufacturing firms to transition technologies from research into manufacturing processes were at the heart of the discussions. Enhancing university-industry collaboration is an important step to help SMEs move new ideas into commercialization. To promote university-industry collaboration, effective research and licensing agreements need to be developed. Participants proposed Phase 0 (seed) Small Business Innovative Research awards and funding for incubators were also discussed as possible incentives to connect U.S. manufacturers to innovation programs created by universities.

C. Education and Workforce Development Workstream

Concerns regarding the lack of worker preparation for high-skilled manufacturing jobs in today's workforce dominated the discussion. Partnerships between community and technical colleges, universities, industry, and government were frequently mentioned as being of central importance to training the next generation of workers. The perceived negative image of manufacturing was also a major theme at each regional meeting. Discussion focused on how to create an image of manufacturing as a stable, socially acceptable career path. Workers and students must be interested and excited about their career prospects to strength the talent pipeline. Participants discussed that repairing this image for all stakeholders, including students, teachers, parents, guidance counselors, essentially all citizens, is critical to maintain and revitalize U.S. manufacturing. To prepare the workforce, community groups and technical colleges need to work with industry to design programs to prepare curricula to ensure that students are learning important job skills. Industry can collaborate with educational institutions by contributing to class projects, and providing cooperative education and internship and externship opportunities. Industry can also participate in programs that educate teachers and parents.

D. Shared Facilities and Infrastructure Workstream

Similar to the work in the technology development workstream, ideas discussed in the breakout sessions on shared facilities and infrastructure evolved over the four regional meetings. Several major themes were common across all the discussions. Many representatives from small and medium enterprises (SMEs) expressed that they were not aware of shared facilities at national laboratories or universities. Participants proposed a database of existing shared facilities that has the potential to allow businesses to identify resources they could use to improve their manufacturing processes and products. Supercomputing centers and centers of excellence where industry could collaborate with universities and develop new technology or prototypes were also listed as potentially useful for businesses.

Centers could be physical or digital, depending on regional technology needs. By the fourth regional meeting, at the University of Michigan, the participants further developed the idea of building private-public manufacturing innovation centers and provided valuable input about how such a center could effectively operate. They discussed ideas on governance, operation, and maintenance of the centers; intellectual property rights of users; and workforce training for the centers. Such centers could help to develop a digital manufacturing infrastructure that would provide businesses support for the design and analysis software to enhance their manufacturing capabilities. Such centers would need to be narrow enough to serve the needs of a particular industry, but broad enough to encourage different aspects of production that cut across industries.

E. Next Steps

The AMP Steering Committee is using the valuable input from regional meetings, the AMP website, and ongoing discussions with stakeholders to generate recommendations for its final report which will be released in spring 2012.

Appendix: Advanced Manufacturing Technology Regional Meeting Agendas and Summaries

Georgia Institute of Technology, October 14, 2011

Time	Event	Speakers/Participants
9:00 am	Welcome	*G.P. "Bud" Peterson* President, Georgia Institute of Technology *Joseph J. Ensor* Vice President and General Manager for Engineering, Manufacturing and Logistics/Electronic Systems Sector, Northrop Grumman Corp. *Jason Miller* Special Assistant to the President for Manufacturing Policy
9:30 am	Overview of Workstreams	*Carrie Houtman* Senior Public Policy Manager, The Dow Chemical Company *Ben Wang* Incoming Executive Director, Manufacturing Research Center; Georgia Institute of Technology/FSU
9:40 am	Government Panel Presentations	*David Hart* (Moderator) Assistant Director for Innovation Policy, Office of Science and Technology Policy *Neal Orringer* Director of Manufacturing and Industrial Base Policy, Department of Defense *Leo Christodoulou* Program Manager, Industrial Technologies Program, Department of Energy *Michael F. Molnar* Chief Manufacturing Officer, National Institute of Standards and Technology *Steven H. McKnight* Director, Civil, Mechanical, and Manufacturing Innovation Division (CMMI) of the Directorate for Engineering, National Science Foundation
10:50 am	Breakout Session Instructions	*Steve Cross* Executive Vice President for Research, Georgia Institute of Technology
11:00 am	Break	
11:15 am	Breakout Session 1	**Technology Development** **Education and Workforce Development** **Shared Facilities and Infrastructure** **Policy**
12:00 pm	Lunch	
12:15 pm	Breakout Session 2	**Technology Development** **Education and Workforce Development** **Shared Facilities and Infrastructure** **Policy**
1:00 pm	Break	
1:15 pm	Breakout Session Outbriefs	
1:50 pm	Next Steps	*Jason Miller* Special Assistant to the President for Manufacturing Policy
2:30 pm	Panel Discussion: Industry Prospective on the AMP Workstreams	*John Zegers* (Facilitator) Director, Georgia Center of Innovation for Manufacturing Panelists: *Walt Stadnisky* President, Roper Pumps *Burl M. Finkelstein* Kason Industries *Donald L. Deptowicz* PCC Airfoils, LLC *David Stern* Senior Vice-President, CardioMEMS, Inc.

Meeting Summary: Georgia Institute of Technology, Atlanta, GA October 14, 2011

Introductory Session

Speakers were G.P. "Bud" Peterson, President of Georgia Institute of Technology; Joseph Ensor, Vice President of Northrop Grumman; and Jason Miller, Special Assistant to the President for Manufacturing Policy. They emphasized the following points:

- Innovation in manufacturing requires a strong manufacturing base, and a strong manufacturing base requires collaboration between the research and industry communities and a robust supply of skilled technical workers.
- Companies are willing to invest to create jobs, but they need to see long-term payoff.
- The goal of the AMP is to accelerate technology development and lower the risk in moving from laboratory to factory to market

Federal Government Panel

Panelists from the government were Leo Christodoulou, Department of Energy (DOE); Steven McKnight, National Science Foundation (NSF); Neal Orringer, Department of Defense (DOD); and Michael Molnar, National Institute of Standards and Technology (NIST). They described initiatives and grant mechanisms aimed at promoting collaborative technology research and technology transfer to industry. Some of the main points included:

- DOE's Innovative Manufacturing Initiative is investing in the next generation processes as well as in the next generation of materials.
- Advanced manufacturing at NSF includes fundamental and cross-cutting research, as well as attention to education and human capital development
- DOD funds the development of high-technology products with long lifetimes. To guarantee the supply chain, DOD monitors and co-invests in the development, purchasing, and maintenance of these high-tech products.
- NIST has many programs that focus on advanced manufacturing, and partners with industry and universities through programs such as the Manufacturing Extension Partnership (MEP) and some shared facilities.

Breakout Sessions Reports

The participants divided into groups, which each discussed the workstream topics of (1) advanced technology development, (2) shared infrastructure and facilities, (3) policy, and (4) education and workforce development. The groups reconvened to offer summaries, including the points described in the following subsections.

Advanced Manufacturing Technology Development

Areas seen as having urgent importance included energy (old and new sources, efficiency, innovative technology); microelectronics; informatics (especially in health care); materials (composites, nanomaterials, and biomaterials) and technology for forming and shaping materials); supply of strategic materials (including rare earth materials); green technology; and broadly applicable systems (automation, supply chain models, predictive modeling, complex systems, and information management.

Shared Facilities and Infrastructure

Participants agreed on the value of shared facilities allowing industry access to costly specialized equipment, but recognized problematic issues such as improving access by small- and medium-sized companies and safeguarding intellectual property. A fundamental need is for a system that lets industry know what facilities are available and what their potential value is in specific manufacturing contexts.

Policy

The burden of regulatory compliance (including domestic regulations and export controls) is an overwhelming issue for many industries, which frankly see Federal regulation as an adversarial bureaucracy. Participants also saw a need for more Federal support of university-industry collaboration; the German model was frequently mentioned in this context.

Education and Workforce Development

Many industries have begun working with schools and community colleges to promote technical education and bolster manufacturing as a career choice. Intervention must be early in students' school careers to be effective. Much of this work is local and needs to be shared and coordinated on a national scale.

Meeting Summary: Massachusetts Institute of Technology, Cambridge, MA November 28, 2011

Introductory Session

Speakers at the opening session were Susan Hockfield, President of MIT and co-chair, AMP Steering Committee; Deval Patrick, Governor of Massachusetts; and Andrew Liveris, Dow Chemical Co and co-chair, AMP Steering Committee, who joined by video. They raised the following points:

- Manufacturing is "central to our national identity," but it "has truly eroded." The erosion of manufacturing threatens the nation's ability to innovate.
- In fall 2010 MIT created a Production in the Innovation Economy (PIE) program that will provide an evidentiary base for policies that link production and innovation
- The Governor announced that Massachusetts is establishing an Advanced Manufacturing Collaborative to "amplify the AMP."
- In the view of many, manufacturing is still dirty, dangerous, low-paying work. A central mission of the new Collaborative is to convey that modern manufacturing drives innovation and offers skilled, high-paying jobs.

Massachusetts Institute of Technology, November 28, 2011

Time	Event Name	Speakers/Participants
10:00 am	Welcome and Overview	*Susan Hockfield* President of MIT The Honorable *Deval Patrick* Governor of Massachusetts Video: *Andrew Liveris* Chairman and CEO, The Dow Chemical Company
10:20 am	Federal Government Panel	*Susan Hockfield* (Moderator) *Subra Suresh* Director of the National Science Foundation *Patrick Gallagher* Under Secretary of Commerce for Standards and Technology, NIST Director *Ken Gabriel* Deputy Director, Defense Advanced Research Projects Agency *Henry Kelly* Acting Assistant Secretary, Office of Energy Efficiency and Renewable Energy
11:30 am	Overview of AMP	**White House Perspective:** *Jason Miller* Special Assistant to the President for Manufacturing Policy **Summary of AMP Workstreams:** *Carrie Houtman* Senior Public Policy Manager, The Dow Chemical Company *Ben Wang* Chief Manufacturing Officer, Georgia Institute of Technology
11:45 am	Breakout Session Instructions	*Martin Schmidt* Associate Provost and Professor of Electrical Engineering, MIT
11:50 am	Break	
12:10 pm	Lunch Breakout Sessions	**Policy:** *Eric Nakajima* (Facilitator) Senior Innovation Advisor, Massachusetts Executive Office of Housing and Economic Development **Education, Training and Recruitment:** *Nancy Snyder* (Facilitator) President and CEO of Commonwealth Corporation **Networks and Shared Facilities:** *Patrick Larkin* (Facilitator) Director of the John Adams Innovation Institute at the Massachusetts Technology Collaborative **University and Industry Collaborations:** *Olivier de Weck* (Facilitator) Associate Professor of Aeronautics and Astronautics and Engineering Systems, MIT; Executive Director, MIT Production in the Innovation Economy (PIE) Study **Technology:** *Ahmed Busnaina* (Facilitator) William Lincoln Smith Professor and Director of the National Science Foundation Nanoscale Science and Engineering Center (NSEC) for High-Rate Nanomanufacturing at Northeastern University **Energy/Sustainability/Green Manufacturing:** *Timothy Gutowski* (Facilitator) Professor of Mechanical Engineering, MIT
1:10 pm	Break	

Time	Event	Speakers/Participants
1:15 pm	Regional Panel 1: Advanced Manufacturing Success Stories	*Karen Mills* (Moderator) Administrator of the U.S. Small Business Administration *Guy Broadbent* President and CEO, Xcellerex, Inc. *Jill Becker* CEO, Cambridge NanoTech *Michael Casper* Founder, President and CEO, UltraSource, Inc. *Joanna Dowling* Director, The Custom Group *Bill Emhiser* President, Maine Manufacturing
2:15 pm	Regional Panel 2: University Activities and Partnerships	*Suzanne Berger* (Moderator) MIT's Raphael Dorman-Helen Starbuck Professor of Political Science and Co-chair of the Production in the Innovation Economy (PIE) project *Mark Trusheim* Bio-manufacturing Executive in Residence, UMASS Dartmouth *Bernhardt Trout* Professor of Chemical Engineering at MIT and Director, Novartis-MIT Center for Continuous Manufacturing *Dean Fuleihan* University of Albany, College of Nanoscale Science and Engineering, Executive Vice President for Strategic Partnerships *Andre Sharon* Professor, Mechanical Engineering at Boston University and Executive Director, Fraunhofer USA Center for Manufacturing Innovation
3:15 pm	Break	
3:30 pm	Regional Panel 3: Regional Government and Policy	*David Hart* (Moderator) Assistant Director of Innovation Policy, Office of Science and Technology Policy *Gregory Bialecki* Secretary of Housing and Economic Development, MA *Lawrence Miller* Secretary, Agency of Commerce and Community Development, VT George Bald Commissioner, Department of Resources and Economic Development, NH *Aaron R. Fichtner* Assistant Commissioner, Labor Planning and Analysis, New Jersey Department of Labor and Workforce Development
4:30 pm	Regional Panel 4: Regional Manufacturing Challenges and Opportunities	*Gururaj (Desh) Deshpande* (Moderator) Chairman of Sparta Group LLC *Marc Giroux* SVP Manufacturing Technology and Engineering, Chief Engineer at Corning *Luis Izquierdo* Vice President of Corporate Operations at Raytheon *Geoff MacKay* President and CEO of Organogenesis *Raymond Stata* Co-Founder of Analog Devices *Daniel Armbrust* President and CEO of SEMATECH
5:40 pm	Closing Remarks	*Susan Hockfield*

Federal Government Panel

Panelists were Subra Suresh, director of NSF; Patrick Gallagher, director of NIST; Ken Gabriel, deputy director of DARPA; and Henry Kelly, acting assistant secretary of DOE. Some of their points:

- Advanced manufacturing is both an enabler of existing products and a source of new products.
- Advanced manufacturing requires a sustained whole-of-government effort. The agencies on this panel will work together to this end.
- While the NSF focuses on basic research, it has long supported manufacturing research, especially through its Engineering Research Centers. New programs, such as Innovation Corps, will enhance NSF's impact on advanced manufacturing
- Regional, long-term public-private partnerships are essential to manufacturing, and NIST, along with the rest of the federal government, wants to facilitate them.
- DARPA's advanced manufacturing programs have as their central focus reducing the time from design to production of manufacturing innovations.
- Advanced manufacturing is at the core of innovation in clean energy technology, making industrial processes more energy efficient and cutting the costs of clean energy products.

Breakout Sessions

Participants divided into six groups to discuss the workstream topics. They reconvened to offer the following points:

- Three primary barriers to investment in US advanced manufacturing, especially for SMEs, are tax/innovation policy, regulatory policy, and trade policy.
- The R&D tax credit should be made permanent.
- We need policies that reward companies for taking technology and business risks.
- The image of manufacturing as "dirty, dangerous, and degrading" is outdated and must be corrected through public education.
- Educational institutions must work with industry to train the future workforce.
- Sharing of infrastructure, facilities and public-private "product accelerators" that help bridge the gap between research/design and production should be supported.
- University/industry collaborations through pre-competitive partnerships, apprenticeships, fee-for-service projects, visiting professorships, and other means are important.
- Cross-cutting technologies, including energy efficiency, modeling and simulation, and advanced sensing and measurement, are required for advanced manufacturing.
- Emerging technologies are vital to US economic development include nanomanufacturing/advanced materials, robotics, custom manufacturing.
- Developing sustainable/green manufacturing depends most prominently on policy.

Regional Panel 1: Advanced Manufacturing Success Stories

Members of this panel were leaders of successful small businesses, most of which make platforms, tools, or systems for advanced manufacturers. All said they could grow faster if they could find more trained staff. Several discussants noted the lack of student interest in science and engineering, especially at community colleges. One noted that some guidance counselors advise students against careers in manufacturing.

University of California, Berkeley, and Stanford University, Berkeley, CA December 5, 2011

Time	Event	Speakers/Participants
8:30 am	Welcome	*S. Shankar Sastry* Dean, College of Engineering, UC Berkeley – Sibley Auditorium, BEC
8:35 am	Opening Remarks and Workstream Overview	*Robert J. Birgeneau* (Workstream Lead) Chancellor, UC Berkeley; AMP Steering Committee *Friedrich B. Prinz* Chairman, Mechanical Engineering Department, Stanford University *Krishna Mikkilineni* Senior Vice President, Honeywell; AMP Workstream Lead *Tom Kalil* Deputy Director for Policy, Office of Science and Technology Policy
9:00 am	Government Panel and Presentations	*Arati Prabhakar* (Moderator) Chair, EERE Advisory Council, U.S. Department of Energy Government Panelists: *David Brinkley* Department of Defense *Leo Christodoulou* Director, Industrial Technologies, Department of Energy *Patrick Gallagher* Director, National Institute of Standards and Technology *Bruce M. Kramer* Senior Advisor, National Science Foundation *Thomas Lee* Director, Microsystems Tech Office, Defense Advanced Research Projects Agency
9:50 am	Introduction to Breakout Session Topics Materials Genome Working Meeting (parallel session)	*Krishna Mikkilineni* (Moderator) Senior Vice President, Honeywell; AMP Workstream Lead Industry Panelists: *Will Coleman* Partner, Mohr Davidow Ventures *Matthew Ganz* Vice President, Boeing *Kurt Petersen* Chief Engineer, Profusa *Dan Jones* Director, Intuitive Surgical *Omkaram Nalamasu* CTO, Applied Materials *Darlene J.S. Solomon* CTO, Agilent Technologies
10:45 am	Breakout Sessions	Clean Energy Cyber-Physical Systems Medical Devices Small-Medium Enterprises Sustainable Manufacturing Synthetic Biology
Noon	Lunch	
1:00 pm	Topline Reports from Breakout Sessions	
1:45 pm	Next Steps	Speaker TBD

Regional Panel 2: University Activities and Partnerships

Members of this panel agreed that universities were good at innovating, but poor at supporting young firms. "We need to make problems of scale-up as interesting to students as those of start-ups," stated one member. The federal government has not been a strong in applied research for advanced manufacturing. The lack of prototyping centers was identified as a key "gap" between research and production.

Regional Panel 3: Regional Government and Policy

This panel featured workforce and economic development leaders from Massachusetts, Vermont, and New Jersey, who noted both the poor image of manufacturing and the low classroom demand for advanced manufacturing-related topics. States need "deeper partnerships" with the federal government, including more flexibility in how they use federal funding and "room to fail once in a while." At a fiscal level, states need more predictability and better access to data.

Regional Panel 4: Regional Manufacturing Challenges and Opportunities

This panel featured representatives of medium-sized and large manufacturers who agreed that manufacturing in a high-tech environment requires close coupling of R&D and customer. U.S. firms cannot be competitive if they are isolationist. For instance, Organogenesis, which manufactures regenerative medicine products, designed its new manufacturing plant so that all functions could be together in the U.S.

Closing quote from Susan Hockfield: "The greatest real thrill that life offers is to create something useful. Too often we fail to recognize and pay tribute to the creative spirit." – Alfred P. Sloan, Jr.

Introductory Session

S. Shankar Sastry, Dean of the UC Berkeley College of Engineering, was joined by Robert Birgeneau, Chancellor of UC Berkeley and AMP Steering Committee; Friedrich B. Prinz, Chairman, Mechanical Engineering Department, Stanford; Krishna Mikkilineni, Honeywell and AMP Workstream Lead; and Tom Kalil, Deputy Director for Policy, OSTP, who joined by video. They raised the following points:

- The nation is on the verge of a new industrial revolution, powered by innovative technologies.
- A goal of the AMP is to establish manufacturing as a platform to revitalize the economy and create more jobs.
- To remain competitive, the nation must re-couple its design and production functions, align manufacturing with policy, and excite today's students.
- The meeting should make concrete proposals, including "some manufacturing moon shots that can motivate and inspire."

Federal Government Panel

Arati Prabhakar (DOE) moderated a federal government panel consisting of Leo Christodoulou (DOE); Patrick Gallagher (NIST); Thomas Lee (DARPA); David Brinkley (DOD); and Bruce Kramer (NSF). Key points included:

- Manufacturing is increasingly about systems: interoperability, supply chains, and information sharing. "Separation is okay; segregation not."
- Create more hubs, networks, clusters, and "shoulder rubbing."
- Physical or virtual hubs and shared facilities/infrastructure enable SMEs to reduce costs and increase access to expensive tooling, characterization facilities, and other platform technologies.
- Fundamental advances in synthetic biology are likely to disrupt manufacturing.
- A sound strategy is to invest in research on process technology that has a wide range of applications itself or leads to products with a wide range of applications.
- The future of manufacturing may be about "satisfying the long tail problem," making a larger number of products in smaller quantities.
- Revitalized manufacturing depends on updating its image as a career and strengthening the workforce—from technicians to PhDs.

Industry Panel

Krishna Mikkilineni (Honeywell) moderated an industry panel consisting of Kurt Petersen (Profusa); Darlene Solomon (Agilent Technologies); Omkaram Nalamasu, (Applied Materials); Will Coleman (Mohr Davidow Ventures); Dan Jones (Intuitive Surgical); and Matthew Ganz (Boeing). The following points were made:

- Synthetic biology is "on the cusp" of designing cells, engineering biological systems, transforming manufacturing, and powering economic growth.
- As companies scale technologically, they will require capital starting with venture capital, to a mix of private equity, corporate funding, partners, and debt financing. Obtaining this continuum of funding can be challenging for companies, especially small companies, so government support through grants and public-private partnerships is also needed.
- Products in some industries, such as aircraft, are too complex for co-location, so "mastering decentralized teamwork is core."
- In China and Singapore, scientists and engineers are "rock stars." This should be a model for the United States, where there is a "huge bias" against manufacturing careers.

Breakout Sessions Reports

The participants divided into six groups, which discussed (1) clean energy, (2) cyber-physical systems, (3) medical technology and devices, (4) sustainable manufacturing, (5) SMEs, and (6) synthetic biology. A parallel session addressed the materials genome. The participants reconvened to offer summaries, including these points:

- Leveraging existing shared facilities, such as national laboratories, and creating new ones are needed to help startups scale their technologies and financing.
- Access to prototyping and virtualization tools can speed technology development.
- Hub/cluster models that co-locate SMEs, universities, community colleges, corporate laboratories, institutes, business incubators, and financing should be promoted.

- Workforce development must include settings where students "get their hands dirty" and learn about life-cycle design; industry must "pull" and clarify its educational needs.
- Outreach and communication are essential to update image of manufacturing.
- The introduction of synthetic biology will spur GMO(genetically modified organisms)- like debates, which need to be discussed early.
- Prompt development of standards and clear, coordinated tax and regulatory policies can open doors to collaboration.
- Many of the groups discussed the need for immigration reform and making the R&D tax credit permanent.

S. Shankar Sastry closed the meeting by recognizing the "passion" of the attendees and urging participants to "stay engaged."

Introductory Remarks

Stephen R. Forrest, Vice President for Research, University of Michigan, moderated an introductory session by Mary Sue Coleman, President, University of Michigan; Theresa Kotanchek, Vice President, Sustainable Technologies and Innovation Sourcing, Dow Chemical Company; Rosina Bierbaum, Professor, University of Michigan School of Natural Resources and the Environment, and Member, PCAST; and Chuck Thorpe, Assistant Director for Advanced Manufacturing and Robotics, OSTP. Some highlights included:

- Michigan has suffered from the decline in manufacturing, but the state is well positioned for a manufacturing resurgence, partly because of its tradition of collaboration.
- Streamlining the path to market requires better models for public-private partnerships (PPPs), such as hubs and clusters.
- Essential to preparation of a skilled workforce are shop courses, positive guidance counseling, and a return to a culture that values working with the hands.

Federal Government Panel

Chuck Thorpe (OSTP) moderated a panel consisting of Michael Molnar (NIST); Neal Orringer (DOD); Leo Christodoulou (DOE); and Steven McKnight (NSF). Some highlights included:

- Advanced manufacturing is ranked "first and foremost at NIST," where all laboratories and most test beds are shared facilities. Funding of $25 million is specifically allocated to advanced manufacturing in response to PCAST.
- Public-private partnerships (PPPs) should be led by industry and university, not government.
- The White House named Department of Commerce Secretary John Bryson and National Economic Council Director Gene Sperling as co-chairs of the White House Office of Manufacturing Policy. The Office of Manufacturing Policy is part of the National Economic Council in the White House and works across Federal government agencies to coordinate the execution of manufacturing programs and the development of manufacturing policy.

University of Michigan, Ann Arbor, MI December 12, 2011

Time	Event	Speakers/Participants
8:30 am	Welcome and Overview	*Stephen R. Forrest* Vice President for Research, University of Michigan *Mary Sue Coleman* President, University of Michigan *Theresa G. Kotanchek* Vice President, Sustainable Technologies and Innovation Sourcing, Dow Chemical Company *Rosina Bierbaum* Professor, University of Michigan School of Natural Resources and the Environment, and Member, President's Council of Advisors on Science and Technology
9:15 am	Overview of AMP Workstreams	*S. Jack Hu* *J. Reid* *Polly Anderson* Professor of Manufacturing Technology, University of Michigan, College of Engineering *Carrie Houtman* Senior Policy Analyst, Dow Chemical Company
9:25 am	Government Panel and Presentations	*Chuck Thorpe* Assistant Director, Advanced Manufacturing and Robotics, Office of Science and Technology Policy *Michael Molnar* Chief Manufacturing Officer, Department of Commerce, National Institute of Standards and Technology *Neal Orringer* Director of Manufacturing, Manufacturing and Industrial Base Policy, Department of Defense *Leo Christodoulou* Program Manager, Advanced Manufacturing Office, Department of Energy *Steven H. McKnight* Director, Division of Civil, Mechanical, and Manufacturing Innovation, Directorate for Engineering, National Science Foundation
10:25 am	Panel Discussion: Industry Perspective on Advanced Manufacturing	*Richard Jarman* (Moderator) President, National Center for Manufacturing Sciences *James P. Tetreault* Vice President, North America Manufacturing, Ford Motor Company *Dawn White* President/CTO, Accio Energy *John Winzeler* President, Winzeler Gear *Sujeet Chand* Chief Technology Officer, Rockwell Automation *Douglas Dinon* Site Leader, Advanced Manufacturing Technology Center in Michigan, General Electric Global Research
11:30 am	Breakout Session Instructions	*S. Jack Hu*
11:35 pm	Lunch	
12:15 pm	Breakout Sessions	**Education and Workforce Development:** *Albert Shih* (U-M) and *Carrie Houtman* (Dow)—Facilitators **Policy:** *Marvin Parnes* (U-M) and *Ed Rozynski* (Stryker)—Facilitators **Shared Infrastructure and Facilities:** *Don Chaffin* (U-M) and *Karen Huber* (Caterpillar)—Facilitators **Technology Development:** *Euisik Yoon* (U-M) and *Theresa G. Kotanchek* (Dow)—Facilitators **Materials Genome Initiative:** *John Allison* (U-M)—Facilitator
1:45 pm	Break	
2:00 pm	Breakout Session Outbriefs	
2:40 pm	Next Steps	*Chuck Thorpe*

- DOD is responsible for about 80% of government manufacturing expenditures. Its main goals are to maximize productivity and drive down costs and delivery times.
- DOE recently created the Advanced Manufacturing Office (AMO), which invests in pervasive, broadly applicable manufacturing processes and next-generation materials. Through partnerships, the AMO will help to develop and deploy new technologies.
- NSF targets PPPs through the Grant Opportunities for Academic Liaison with Industry (GOALI), Engineering Research Centers (ERC), Industry-University Cooperative Research Centers (I/UCRC), and the Innovation Corps (i-Corps) programs. These NSF grants encourage academics to collaborate with industry and, in some cases, to gain industry experience.

Industry Perspective on Advanced Manufacturing

Richard Jarman (National Center for Manufacturing Sciences) moderated a panel of Sujeet Chand (Rockwell Automation); James P. Tetreault (Ford Motor Company); Dawn White (Accio Energy); John Winzeler (Winzeler Gear); and Douglas Dinon (General Electric Global Research). Some key points included:

- Innovation occurs in PPPs, which bring together talent, investment, and infrastructure. Manufacturing issues should not be divided into concerns of SMEs vs. small firms; "it's really an ecosystem" of PPPs. For General Electric, "the public-private partnership is our approach going forward."
- The supply chain model offers "huge opportunities" to optimize end-to-end processes.
- Lack of expertise in something as basic as joining dissimilar materials for automobile applications "should concern us;" These skills are often obtained abroad. Several other countries have national technical education curricula; by age 16, "skills are extraordinarily high."
- A small firm making plastic gears succeeds through "true strategic partnerships" with a university, small and large firms, outreach to high school counselors, and community involvement.
- Some small firms gain great benefits from national laboratories, but access is difficult.
- Suspend tax on repatriating overseas cash, incentivize to rebuild infrastructure. Make R&D tax credit permanent and offer tax credit for training scientists and engineers.

Breakout Sessions Reports

The participants divided into five groups, which discussed the four workstream topics plus the Materials Genome Initiative. They reconvened to offer suggestions, including the following:

- Ensure that regulations are not barriers to entry for small and medium firms.
- Consider policy measures such as a domestic manufacturing deduction, an R&D tax credit for research done off-site (for example, working with a university), and a tax credit for new factories or for fighting intellectual property theft.

- Use master agreements, not ad hoc bargaining, for university-industry intellectual property rights discussions.
- In technology development, emphasize the importance of adhesives, joining, and fastening technologies, as well as non-destructive evaluation, sustainable manufacturing, and additive manufacturing.
- Advance the optimization of the supply chain across several suppliers, not just between one supplier and one customer.
- In sharing facilities, use physical hubs, digital hubs, and open source databases. The main purpose of these hubs and databases is to connect small and medium firms with resources, partners, and the community.
- In education, better align the image of manufacturing with interests of young people; better align academic output with the needs of industry; reduce barriers to using veterans' skills, for example, by translating military occupational classes to civil categories.

Next Steps

- Chuck Thorpe of the Office of Science and Technology Policy concluded the meeting with a call for participation, and a sketch of future objectives. The AMP has now been discussed by well over 1,000 participants at the four regional meetings.
- The continuing involvement of those participants is needed to prioritize action plans and generate recommendations in spring 2012.
- After the report comes the real work of "action and evangelism," The goal is to spread the word, developing action plans, and forming new partnerships.

In: Advanced Manufacturing in America
Editors: Brandon Cullerne and Michael Laston
ISBN: 978-1-62257-932-7

Chapter 9

THE OBAMA ADMINISTRATION'S PROPOSAL TO ESTABLISH A NATIONAL NETWORK FOR MANUFACTURING INNOVATION*

John F. Sargent Jr.

ABSTRACT

Manufacturing plays an important role in the nation's economy, employment, and national defense. Accordingly, Congress has maintained a strong interest in the health of the U.S. manufacturing sector. Some analysts have expressed concerns about a decades-long decline in manufacturing employment punctuated by a steep drop from 2001 to 2010, as well as about the offshore outsourcing of production and related functions, such as research and development, by U.S. manufacturers. Others see the U.S. manufacturing sector as vibrant and healthy as evidenced by growth in output and productivity.

The Obama Administration has undertaken a number of initiatives intended to support U.S. manufacturing, including establishment of the Advanced Manufacturing Partnership, Advanced Manufacturing National Program Office (AMNPO), Advanced Manufacturing Technology Consortia program, National Robotics Initiative, and Materials Genome Initiative. In his FY2013 budget, President Obama proposed the creation of a National Network for Manufacturing Innovation (NNMI) to help accelerate innovation by investing in industrially relevant manufacturing technologies with broad applications, and to support manufacturing technology commercialization by bridging the gap between the laboratory and the market. The NNMI proposal calls for the establishment of up to 15 Institutes for Manufacturing Innovation (IMI) funded through a one-time infusion of $1 billion in mandatory funding to the Department of Commerce's National Institute for Standards and Technology (NIST) and carried out over a period of 10 years. Each IMI would be comprised of stakeholders from industry (including large companies and small- and medium-sized manufacturing enterprises), academia, federal agencies, and state government entities. According to the proposal, each IMI is to be competitively selected, serve as a regional hub for manufacturing innovation (as well as part of the national network), and have a unique focus area (e.g., an advanced material,

* This is an edited, reformatted and augmented version of Congressional Research Service, Publication No. R42625, dated July 25, 2012.

manufacturing process, enabling technology, or industry sector). The NNMI would be managed collaboratively by NIST, the Department of Defense, Department of Energy, National Science Foundation, and other agencies. An Administration official has testified that the NNMI program has not yet begun and that the establishment of the NNMI will require congressional authorization and funding. No legislation has been introduced to establish the NNMI. However, the President announced his intention to establish a pilot institute using current year funding; in this regard, the Department of Defense issued a Broad Agency Announcement (BAA) in May 2012 to award up to $30 million for an institute focused on additive manufacturing. The pilot is described as "the first institute to be launched within the National Network for Manufacturing Innovation" and "a proof-of-concept for the potential subsequent institutes." The BAA limits the types of organizations that can compete for the award of this institute to universities and non-profit organizations. Universities must incorporate in their proposals a partnership with a cluster of manufacturing firms and associated institutes; non-profit industry consortia must incorporate a partnership with universities for applied research. The AMNPO has published a request for information seeking public comment on various aspects of the NNMI to aid in the development of the institutes and the network. Topics included in the RFI include technology focus areas, structure, governance, sustainability, and education and workforce development.

Overview

Congress maintains a strong interest in the health of the U.S. manufacturing sector due to its central roles in the U.S. economy and national defense. With respect to the economy, manufacturing accounts for about 12% of the nation's gross domestic product (GDP) and nearly two-thirds of U.S. exports. Manufacturing enterprises directly employ nearly 12 million U.S. workers and indirectly support millions of additional jobs in other industries (e.g., banking, shipping, insurance). Manufacturers also fund about two-thirds of the nation's industrial research and development (R&D), providing a foundation for technological innovation and continued U.S. technological leadership. In addition, manufacturing workers earn higher annual pay and benefits (an average of $77,186 in 2010) than their non-manufacturing counterparts (an average of $56,436 annually).[1] With respect to national defense, the United States depends heavily on its manufacturing base to produce the weapons, aircraft, vehicles, ships, and other equipment needed to protect the nation.

Analysts hold divergent views of the health of U.S. manufacturing. Some see the U.S. manufacturing sector as vibrant and healthy. Those holding this view tend to point to, among other things, the sector's strong growth in output and productivity, as well as the United States' world-leading share of global manufacturing output. In addition, some analysts note that between January 2010 and June 2012 manufacturing employment added about half a million jobs, growing to nearly 12.0 million.[2,3] Other analysts believe that the U.S. manufacturing sector is at risk. Among the expressed concerns of those holding this view, include:

- a "hollowing-out" of U.S. manufacturing resulting from the decision of many U.S. manufacturers to offshore production activities and other corporate functions (e.g., research and development, accounting, information technology, tax planning, legal research);[4]

- focused efforts by other nations to grow the size, diversity, and technological prowess of their manufacturing capabilities, and to attract manufacturing operations of U.S.-headquartered multinational companies using a variety of policy tools (e.g., tax holidays, worker training incentives, market access, access to rare earth minerals); and
- a decades-long decline in U.S. manufacturing employment, punctuated by a steeper drop from 2001 to 2010. In January 2010, U.S. manufacturing employment fell to its lowest level (11.5 million) since March 1941, down more than 41% from its peak of 19.6 million in June 1979.[5]

The recent recession, relatively slow pace of recovery, and concerns about the prospects for double-dip recession[6] have contributed to increased concerns about the health of U.S. manufacturing. Some stakeholders and policymakers advocate for macro-level changes to improve the business environment, including reducing tax and regulatory burdens on manufacturers and reforming the nation's tort laws.

Others—including President Obama—support more direct and focused efforts funded by the federal government. In particular, President Obama has undertaken and proposed a variety of initiatives (e.g., the Advanced Manufacturing Partnership, the National Robotics Initiative, Materials Genome Initiative) to help address concerns about U.S. manufacturing. One of the President's key proposals to help U.S. manufacturers is the establishment of a National Network for Manufacturing Innovation.

In February 2012, the Obama Administration released A National Strategic Plan for Advanced Manufacturing, a report by the President's National Science and Technology Council, putting forth a strategy to guide federal advanced manufacturing R&D investments. The report notes

> The acceleration of innovation for advanced manufacturing requires bridging a number of gaps in the present U.S. innovation system, particularly the gap between R&D activities and the deployment of technological innovations in domestic production of goods.[7]

The proposed NNMI seeks, in part, to bridge the innovation gap asserted in this report.

ADMINISTRATION PROPOSAL

President Obama proposed the establishment of a National Network for Manufacturing Innovation (NNMI) in his FY2013 budget, and formally introduced the concept on March 9, 2012, in a speech at the Rolls-Royce Crosspointe jet engine disc manufacturing facility in Prince George County, VA. According to the proposal, the purpose of the NNMI is to bring together industry, universities and community colleges, federal agencies, and regional and state organizations

> to accelerate innovation by investing in industrially relevant manufacturing technologies with broad applications, and to support manufacturing technology commercialization by bridging the gap between the laboratory and the market.[8]

In particular, the NNMI seeks to "advance technological innovation at a pace much faster than any one company could on its own,"[9] integrate innovation resources, improve the competitiveness of U.S. manufacturing, and encourage investment in the United States.[10]

The NNMI would consist of

> a network of institutes where researchers, companies, and entrepreneurs can come together to develop new manufacturing technologies with broad applications. Each institute would have a unique technology focus. These institutes will help support an ecosystem of manufacturing activity in local areas. The Manufacturing Innovation Institutes would support manufacturing technology commercialization by helping to bridge the gap from the laboratory to the market and address core gaps in scaling manufacturing process technologies.[11]

The NNMI would be managed collaboratively by the Department of Commerce's (DOC's) National Institute of Standards and Technology (NIST), the Department of Defense (DOD), the Department of Energy (DOE), the National Science Foundation (NSF), and other agencies.[12]

Funding

As proposed in the President's FY2013 budget, NIST would receive a one-time infusion of $1 billion in mandatory funding in FY2013 to be spent over 10 years (FY2013-FY2022, see **Table 1**).[13] Federal funds would be used to help establish and support up to 15 Institutes for Manufacturing Innovation (IMIs, which collectively would form the NNMI) on a cost-shared basis with industrial, academic, and state and local organization partners. Each IMI is expected to become financially sustainable within seven years.[14] Funding for the program would be front-loaded with $206 million in spending projected for FY2013, and a total of $839 million in spending in the first five years.[15]

Table 1. Proposed Schedule of NNMI Expenditures in millions of dollars

FY 2013	FY 2014	FY 2015	FY 2016	FY 2017	FY 2018	FY 2019	FY 2020	FY 2021	FY 2022
$206	$131	$174	$189	$139	$69	$44	$28	$16	$4

Source: Office of Management and Budget, Executive Office of the President, Fiscal Year 2013 Budget of the U.S. Government, February 2012, Table S-9, p. 217.

Structure and Guiding Principles

The National Network for Manufacturing Innovation would be composed of competitively selected, independently managed Institutes for Manufacturing Innovation. Each IMI would have a specific focus area and serve as a regional innovation hub. Focus areas could include an advanced material (e.g., carbon nanotubes), manufacturing process (e.g., additive manufacturing), enabling technology (e.g., nanotechnology), or industry sector (e.g., medical devices).

The institutes would bring together large companies, small and medium-sized enterprises, academia, federal agencies, and state governments to accelerate innovation through co-investment in industrially relevant manufacturing technologies with broad applications. The institutes would be focused on helping to bridge the gap between basic research and product development, provide companies with access to cutting edge capabilities and equipment, and an environment for educating students and training workers in advanced manufacturing skills. The institutes would seek to reduce the costs and risks of commercializing new technologies and to address relevant manufacturing challenges on a production-level scale.[16] While the IMIs are intended to serve as regional hubs for manufacturing innovation in specific focus areas, collectively the institutes would also function as a network for the sharing of knowledge and best practices.

Patrick Gallagher, Under Secretary of Commerce for Standards and Technology/Director of NIST, testifying before the House Committee on Science, Space, and Technology, Subcommittee on Technology and Innovation, articulated a number of key principles that would guide the governance and work of the NNMI:

- An interagency program management team would define the NNMI's and IMIs' organizational design, manage an open and competitive selection process, and execute the awards process. The team would also define the selection criteria to be used, incorporating public input (see "Outreach Efforts").
- Each IMI would integrate capabilities and facilities needed to address crosscutting manufacturing challenges that, if met, have the potential to retain or expand domestic manufacturing on an economically sound basis.
- IMIs would conduct applied R&D and development projects to reduce the cost and risk of commercializing new technologies or solve generic industrial problems, conduct education and training efforts at all levels, develop methodologies and practices for supply chain integration, and engage with small and medium-sized enterprises.
- Each IMI would have a core of two or more companies, incorporate industry in agenda development, and have direct involvement of industry scientists and engineers in institute projects.
- IMI awards would be made in the form of grants, contracts, and cooperative agreements, possibly over multiple rounds of competitions.
- IMI proposers would be expected to show how federal investments would stimulate investments by the organizations comprising the partnership and/or from other non-federal sources.
- Subsequent federal support for an IMI would be contingent on demonstration of this additional investment, on progress to self-sustainability, and on progress toward meeting the goals of the NNMI.[17]

The following additional IMI principles were included in a Request for Information (RFI) published by the Advanced Manufacturing National Program Office (AMNPO) in the *Federal Register*:[18]

- Long-term partnership between industry (including small, medium, and large firms); educational institutions; non-government organizations; and state, regional, and local economic development authorities;
- Flexibility to form integrated teams of industrial and academic experts from multiple disciplines to solve difficult problems and to develop the future workforce;
- Adaptability for education and workforce development at multiple levels, including K-12, professional credentialing, undergraduate and graduate education, and mentoring and professional development;
- Involvement of industry associations, professional societies, and economic development organizations for validation and linkages to broader industry and regional activities;
- Analytical capability to identify critical emerging technologies with transformational impact and operational capacity in translating these technologies into products and businesses for the market;
- Ability to engage and assist SMEs to effectively deploy technologies; and
- A sustained focus on innovation with a strong reputation for quality and success.[19]

The Fraunhofer-Gesellshaft Model

The Fraunhofer-Gesellschaft* is a German-based, application-oriented research organization that seeks to bridge the innovation gap that can exist between fundamental research activities and commercial applications. In particular, Fraunhofer illustrates the position of its work along the innovation path as being between the Germany's Max-Planck Institutes, which conduct fundamental basic research, and industry, and its sources of funding as being balanced between public and private sources. (See graphic below.)

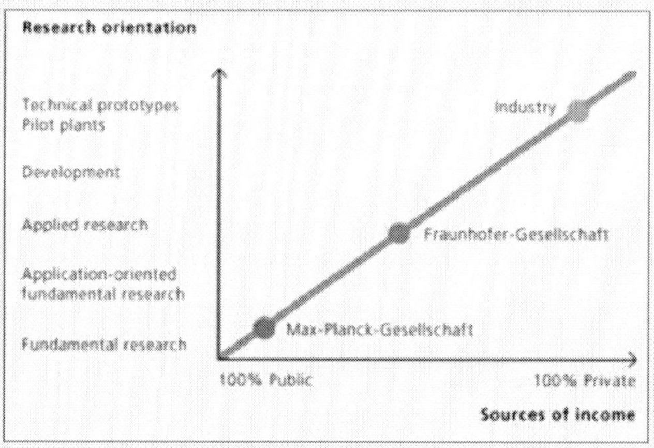

Fraunhofer has more than 80 research units, including 60 Fraunhofer Institutes, throughout Germany with more than 20,000 staff members. Fraunhofer reports its annual research budget as €1.8 billion (approximately $2.2 billion). More than 70% of Fraunhofer's funding is provided through contract research with industry and publicly financed research projects; the balance is provided by the German federal and Länder (state) governments as base funding used to support longer-term, pre-competitive. According to Fraunhofer, the organization bolsters the competitive strength of the economy in their region, and throughout Germany and Europe, by promoting innovation, strengthening the technology base, improving the acceptance of new technologies, and helping to train scientists and engineers.

Fraunhofer has three subsidiaries, including Fraunhofer USA, a wholly-owned non-profit organization headquartered in Plymouth, Michigan. As of 2010, Fraunhofer USA operated six centers in the United States, each working in collaboration with one or more Fraunhofer Institutes in Germany. Total 2010 revenues for Fraunhofer USA were $40 million, of which $20 million were posted by the Fraunhofer Center for Molecular Biotechnology in Delaware, and $4 million by the Fraunhofer Center for Sustainable Energy Systems in Massachusetts.

* "Gesellschaft" is a German word meaning association, society, or institute.

Sources: Fraunhofer-Gesellschaft website, http://www.fraunhofer.de; Fraunhofer *Annual Report* 2010.

The NNMI is said to be modeled after the German Fraunhofer Institutes (see "The FraunhoferGesellschaft Model" box below), which some consider to be a key facet of Germany's high-tech manufacturing success.[20] The Council on Competitiveness,[21] Information Technology and Innovation Foundation,[22] and President's Council of Advisors on Science and Technology[23] and other organizations have endorsed the NNMI concept or proposed a network of U.S.-based public-private manufacturing centers similar to the NNMI.

PRELIMINARY ACTIVITIES

The Obama Administration has undertaken efforts to lay a foundation for the NNMI in advance of possible congressional authorization and funding. In particular, the Administration is proceeding with the establishment of a pilot institute and has initiated an outreach effort to incorporate the perspectives of industry, academia, and non-profit organizations. In particular, the outreach effort is intended to identify a wide-ranging set of technology focus areas for the IMIs, and ideas on IMI and NNMI institutional design and governance to include issues such as treatment of intellectual property and management structure.[24]

Pilot Institute

In his announcement of the NNMI in March 2012, President Obama also committed to the establishment of a pilot institute using existing resources from the Department of Defense and other federal agencies.[25] In May 2012, DOD published a broad agency announcement (BAA, a tool used for contracting) soliciting technical and cost proposals for an Additive Manufacturing Innovation Institute, describing it as "the first institute to be launched within the National Network for Manufacturing Innovation" and "a proof-of-concept for the potential subsequent institutes."[26] (The BAA describes additive manufacturing as "a revolutionary suite of manufacturing technologies for building up parts, and potentially entire systems, in a layer-bylayer fashion, placing material precisely as directed by a 3D digital file.") The lead agency to launch this institute is the Office of the Secretary of Defense (OSD), Manufacturing and Industrial Base Policy, through OSD Manufacturing Technology. The pilot institute award and management will be a cross-agency effort, primarily led by the Defense-wide Manufacturing Science and Technology Program Office, executed through the Air Force Research Laboratory.

According to the BAA, the pilot initiative, called the Additive Manufacturing Innovation Institute, is intended to

> bridge the gap between basic research and product development, provide shared assets to help companies access cutting-edge capabilities and equipment, and create an unparalleled environment to educate and train students and workers in advanced manufacturing skills. The pilot Institute will serve as a technical center of excellence, providing the innovation infrastructure to support manufacturing enterprises of all sizes.... The goal of the pilot Institute is to increase the successful transition of additive manufacturing technology through advanced manufacturing innovation, create an adaptive workforce capable of meeting industry needs, further increasing

domestic competitiveness, and meet DOD, DOE, and other participating civilian agency requirements.[27]

According to the BAA, total funding for the Additive Manufacturing Innovation Institute is anticipated to be approximately $60 million, of which $30 million would be federal funding provided by several agencies[28] and an additional $30 million which "is strongly desired to be provided as cost share, both direct and in-kind," by the stakeholders.[29] In addition, once the institute is established it would be a "likely candidate for additional funds on a competitive basis"—up to $15 million in federal funding for specific projects.[30] These additional funds bring total potential federal funding to $45 million, and total institute funding to as much as $75 million.

The BAA limits the competition for the pilot institute to universities and non-profit organizations (offerors) that are registered as a U.S. organization.[31] If the offeror is a university, it must propose a partnership with a cluster of manufacturing firms and associated institutes. If the offeror is a non-profit industry consortium, it must propose a partnership with universities for applied research. The BAA encourages the inclusion of small businesses in all offers as part of a teaming arrangement. Federally funded Research and Development Centers (FFRDCs) may be eligible offerors, subject to Federal Acquisition Regulation provisions governing FFRDCs, as well as to other conditions.

An offeror must, among other things, provide the outline of a comprehensive financial plan to achieve sustainability of the institute within five years; provide a detailed business plan that includes appropriate roles for resources from the federal government, state government(s), local government(s), industry, academia, and other formal partners; provide a plan for how other federal programs (such as the Small Business Innovation Research program[32] and Manufacturing Extension Partnership)[33] can be leveraged and used to supplement the institute; and demonstrate that their proposal has the potential to advance manufacturing within the United States.[34]

The NNMI interagency team hosted a proposers' day to share its views of the pilot institute, answer questions, and provide an opportunity for networking among potential team partners. Nearly 300 people from industry, academia, government, and other organizations attended this event.[35]

Outreach Efforts

The Advanced Manufacturing National Program Office has initiated efforts to engage manufacturing innovation stakeholders in industry, academia, and regional and state communities. On May 4, 2012, AMNPO issued a Request for Information (RFI) in the *Federal Register* seeking public comment on various aspects of the NNMI, including technology focus areas, structure, governance, sustainability, and education and workforce development.[36]

In addition, the AMNPO, in cooperation with stakeholders and local organizations, is hosting a series of workshops around the country seeking public comments on the NNMI from SMEs, large manufacturers, universities and community colleges, state and local governments, economic development organizations, and other stakeholders. The first of these workshops was held on April 25, 2012, at Rensselear Polytechnic Institute (RPI) in Troy, NY.

NIST advertised the workshop on its website as an "opportunity to help design the National Network for Manufacturing Innovation."[37] A second workshop was held on July 9, 2012, at the Cuyahoga Community College in Warrensville Heights, OH. Both workshops were organized around the four broad topic areas (and associated questions) published in the NNMI RFI, namely technologies with a broad impact, institute structure and governance, strategies for sustainable institute operations, and education and workforce development.

Over 250 participants attended the first workshop at RPI (approximately 44% from academia; 34% from industry; 10% from economic development organizations; and 12% from local, state, and federal government), offering 1,600 individual ideas. A report, *Designing for Impact I: Workshop on Building the National Network for Manufacturing Innovation: Summary of Results*, summarizing the ideas put forth by stakeholders at the workshop was published in June 2012 and is available on the manufacturing.gov website.[38]

Legislative Status

The creation of the NNMI depends on congressional authorization and funding. The Administration is currently seeking one or more Members in each legislative chamber to work with to develop, sponsor, and champion legislation to authorize and fund this effort.[39] According to the President's proposal, funding for the NNMI would be made through one-time mandatory funding of $1 billion. The use of mandatory funding for the NNMI would allow funding to be provided by a law other than an appropriations bill, removing the funding decision from the regular appropriations process.[40] To date no legislation has been introduced in either chamber.

Issues for Consideration

The proposed National Network for Manufacturing Innovation raises a variety of issues for Congress, some of which were raised in a hearing on the proposal held by the House Committee on Science, Space, and Technology's Subcommittee on Technology and Innovation in May 2012. Among the questions Congress may wish to consider as it takes up legislation to establish and fund the NNMI:

- Is the U.S. global competitive position in manufacturing in decline? If so, what factors are contributing to the decline? What are the potential implications of U.S. manufacturing competitiveness decline?
- Some assert that U.S. manufacturing is healthy and growing, pointing to indicators such as increased output and productivity; others assert that U.S. manufacturing is in decline, pointing to decreased manufacturing employment and the movement of production and related functions to other countries.
- Those that see U.S. manufacturing as healthy generally assert that increased globalization and efforts to facilitate trade naturally lead companies seeking to maximize profits, open new markets, increase global market share, and better serve their customers to locate some of their production and related activities outside the

United States. They also argue that, in competing against other multinational corporations, they must undertake such efforts to remain competitive. Some also argue that a variety of factors in the U.S. market (e.g., tax rates, regulations, tort law) place a heavy burden on U.S. manufacturing; in contrast, other nations may have much lower labor costs and their governments may offer a variety of incentives (e.g., tax holidays, worker training, rapid permitting) to attract and retain the manufacturing and related activities of U.S.-based companies.

- Many who see U.S. manufacturing in decline assert that U.S. manufacturing capacity is being "hollowed" out as production facilities and supporting functions (such as R&D, information technology, and accounting) are sited overseas, leaving only a shell of a corporation located physically in the United States. Some experts assert that a nation's (or a state's or a region's) manufacturing strength depends on a "critical mass" of companies that are engaged in similar and supporting activities, creating a synergy that increases the overall strength of the firms in the cluster due to a number of positive reinforcing factors (e.g., knowledge sharing, attraction of workforce talent, new start-ups, establishment of new plants, co-location of supply chains, improvements in infrastructure). As U.S.-based firms move production and related activities outside the United States, some believe that this "critical mass" may be lost, starting a downward spiral in which the synergies are lost and firms opt to move operations outside the United States where such clusters have developed.
- Accordingly, many who subscribe to this view believe that manufacturing employment will decline, R&D activities will relocate to be near production facilities, and service firms that support manufacturing will be lost. In addition, some assert that the "hollowing out" will result in the loss of manufacturing capabilities needed to support the nation's military and increase the manufacturing know-how of potential adversaries.
- Should the federal government directly or indirectly support the competitive position of the U.S. manufacturing sector? Which federal policies and programs should be prioritized? Why should the NNMI be prioritized over other approaches? Is the NNMI duplicative of other federal efforts?
- There are many views as to what the federal government can and should do to support the competitive position of the U.S. manufacturing sector. In general, some prefer an approach that reduces costs and other burdens on manufacturers, such as reducing taxes, regulations, and frivolous lawsuits. Others prefer an expanded direct role for the federal government, such as increasing federal support for manufacturing R&D, providing grants and loan guarantees for domestic manufacturing, and, in some cases, subsidizing production of products for which there are deemed positive benefits for the nation that cannot be captured by the manufacturer (economists refer to such benefits as positive externalities). With a range of options that might be pursued to improve the competitive position and strength of U.S. manufacturing, some may believe that the NNMI should be given high priority due to its perceived benefits (e.g., advancing research discoveries toward market-ready technologies).
- Others may believe that the role envisioned for the NNMI should be performed by the private sector; that the federal government should not favor or subsidize particular companies, industries, or technologies; that the NNMI would be ineffective or counterproductive; that the funds that would go to the NNMI should be used to

support manufacturing in other ways; that the funds should be used for different federal functions altogether; or that the funds should be directed toward deficit reduction. Some may believe that the NNMI is, in part or in whole, duplicative of other federal programs, such as the Manufacturing Extension Partnership; or, as a new and separate program, represents an increasing fragmentation of federal efforts to help manufacturers.

- How would the "national network" aspect of the NNMI operate? What role would the federal government have in the national network?
- Implicit in the title of the National Network for Manufacturing Innovation is the notion that there would be a national network. The Administration's proposal does not discuss how the individual institutes would function as a network. Congress may want to explore what form the network would take (e.g., a national network office with staff, a database of information), what functions it would perform (e.g., sharing of lessons learned, referrals of companies to centers with specific expertise), who would perform the functions (e.g., NIST, AMNPO, IMIs), and how the performance of these functions would be paid for (e.g., federal funding, private funding from the IMIs).
- Should the NNMI be funded on a mandatory or discretionary funding basis? Should a one-time advance appropriation be provided for the proposed 10-year life of the program, or should the NNMI be subject to annual review, oversight, and consideration in the regular annual appropriations process? Which programs would be cut or eliminated as offsets for the NNMI's proposed $1 billion mandatory appropriation?
- The NNMI could be supported through either mandatory[41] or discretionary funding. The Obama Administration has proposed one-time mandatory funding of $1 billion to be used over the course of 10 years. This approach may provide a higher degree of certainty about the availability of out-year funding for the NNMI (though Congress could opt later to rescind all or part of such funding). Such an approach would require offsets from other mandatory funding. Alternatively, providing funding through annual appropriations might allow Congress greater oversight opportunities and flexibility in modifying the program and its funding levels. If the NNMI were to be supported through discretionary funding, cuts would need to be made from other discretionary spending. The Obama Administration's FY2013 budget request specified a number of cuts in mandatory spending, but did not specify which cuts would be used to offset proposed funding for the NNMI.
- What is the appropriate role of the federal government in manufacturing-related innovation? Should the federal government's role end with basic research funding, or include funding for applied research, development efforts focused on cost reduction and technical feasibility, or demonstration projects?
- There are many views regarding the appropriate role of the federal government in the innovation process. While there has been a general consensus on the federal government's support for basic research, congressional efforts to provide later stage support for innovation (aside from meeting government mission requirements, such as national defense) have been met with opposition from different quarters.
- Among the arguments put forth by supporters of later-stage federal investments (e.g., applied research, development efforts, and demonstration projects):

- In some cases, important benefits—e.g., economic, social, national security— may be achieved that would not otherwise be achieved due to factors such as the absence of market incentives (e.g., development of drugs for diseases or conditions that only affect a small number of people) or the inability of a single company or group of companies to undertake such efforts due to high cost, high risk, and/or a long time horizon for achieving a return on investment. Benefits that cannot be captured by a company (or group of companies working together) that brings a product or service to market are referred to by economists as "positive externalities." Unable to capture these benefits, a company is not likely to consider them in its decision-making regarding whether to pursue the development of such a technology or product. Accordingly, economists assert that the result may be private sector underinvestment in beneficial R&D. Some analysts argue that, in such cases, public investment may be justified to induce the development of these technologies (and the realization of these benefits) by sharing costs and risks.
- Such efforts are needed to ensure U.S. leadership in technologies and industries critical to U.S. national security and economic security.
- Such efforts are needed to offset the industrial policies of other nations that make the business environment in the United States comparatively less attractive.

Among the arguments put forward by opponents of such efforts:

- Such efforts constitute an "industrial policy," resulting in distortions in markets and flows of capital by substituting governmental preferences for market forces.
- Government is not able to make better decisions than markets, therefore federal funding for such activities is generally inefficient or wasted.
- Such efforts constitute "corporate welfare" as they provide direct or indirect public subsidies to for-profit corporations, enriching shareholders and others at the expense of taxpayers.
- Government funding and tax, regulatory, or policy decisions may be used for "crony capitalism," i.e., to explicitly reward political supporters or punish opponents.
- The ability to provide direct federal funding to companies or industries can lead to governmental corruption, fraud, and graft.
- The absence of express authority in the Constitution to engage in such activities makes such efforts unconstitutional.
- The current economic condition of the United States with respect to the budget deficit, national debt, and future financial liabilities does not allow for such expenditures, irrespective of merit and efficacy.
- How can the NNMI contribute to the retention of manufacturing-related activities in the United States, both broadly as well as with respect to the R&D that the IMIs advance toward commercialization?
- The innovation process can be extremely challenging. Even good ideas can fail due to a number of reasons (e.g., technical, cost, and risk barriers; disconnection from market needs; absence of standards; regulatory hurdles). A major thrust of U.S. science and technology policy has focused on how to move new ideas and insights from the laboratory into the marketplace.

- In the past, the strength of the U.S. economy and its position in the global economy largely meant that the success of U.S.-based companies in overcoming the obstacles to innovation and moving a process or product into the market resulted in production-related activities and jobs in the United States. Today, however, companies have increased options (and sometimes incentives) to establish production facilities outside the United States. If a primary goal of the NNMI is to stimulate domestic production activities and to capture the associated jobs, Congress may wish to explore whether and how the NNMI would serve to encourage domestic production based on the technologies it advances through the innovation process.
- Which agency/agencies should lead and manage the NNMI?
- President Obama has proposed that funding for the NNMI be given to the National Institute of Standards and Technology, an agency of the U.S. Department of Commerce, through onetime mandatory funding of $1 billion to be spent over 10 years. The NNMI would be managed collaboratively by the Department of Commerce's (DOC's) National Institute of Standards and Technology (NIST), the Department of Defense (DOD), the Department of Energy (DOE), the National Science Foundation (NSF), and other agencies.
- As the recipient of the funds, NIST appears to be the de facto lead on the initiative. In addition, the Under Secretary for Standards and Technology and Director of NIST was the only Administration official to testify at a House hearing on the NNMI.
- NIST has played an important role in the federal government's efforts to support U.S. manufacturing since its establishment as the National Bureau of Standards in 1901. Beyond its core mission in measurement science (i.e., metrology) and standards, NIST took on its current name and was given additional authorities and programs by Congress in the late 1980s. Among these new programs were several programs focused on supporting U.S. firms, including the Advanced Technology Program (ATP, a program to accelerate the development of generic, pre-competitive, high-risk, high-payoff technologies; ATP was eliminated in 2007 and replaced by the Technology Innovation Program (TIP), which was subsequently eliminated);[42] the Malcolm Baldrige National Quality Award (a program to encourage the adoption of quality management principles by private companies and non-profit organizations); and the Manufacturing Extension Partnership (MEP, a program to enhance productivity, technological performance, and strengthen the global competitiveness of small and medium-sized U.S.-based manufacturing firms).[43]
- The Department of Defense is also playing a key role in the NNMI. In particular, DOD has issued a solicitation to establish a pilot institute focused on additive manufacturing, describing it as "the first institute to be launched within the National Network for Manufacturing Innovation" and "a proof-of-concept for the potential subsequent institutes."[44]
- Congress may wish to consider whether to designate a lead agency for the NNMI and to provide that agency with governance authorities and responsibilities, or whether to parse NNMI authorities and responsibilities among several agencies. Alternatively, Congress might opt to provide the Administration with flexibility to assign agency roles and responsibilities.

- What requirements must be met for an !M! to be considered self-sustaining? With respect to federal policy, what would be the consequences for an !M! that does not become self-sustaining?
- As articulated by the Obama Administration, an IMI is to become self-sustaining no later than seven years from its award date. However, no definition of self-sustaining has been put forward by the Administration, nor is there any indication of consequences for failure to become self-sustaining. In addition, the Administration has stated that it expects that IMIs would be eligible to compete for funds for project specific activities under other (i.e., nonNNMI) federal programs.[45] How would such funding be considered with respect to the requirement for self-sustainability?
- Federal requirements that organizations become self-sufficient have not always met with success. For example, the Omnibus Trade and Competitiveness Act of 1988 (P.L. 100-418), which established the NIST Manufacturing Extension Partnership program,[46] required centers to become self-sufficient within six years:

> In no event shall funding for a Center be provided by the Department of Commerce after the sixth year of the operation of a Center.[47]

Congress later amended the MEP authorities in the Technology Administration Act of 1998 (P.L. 105-309) to allow centers to continue receiving federal funds if the center "has received a positive evaluation through an independent review," though it restricted funding after the sixth year to be no more than "one third of the capital and annual operating and maintenance costs of the Center."[48] Congress may wish to consider whether to legislatively require that IMIs become self-sufficient, the time period within which self-sufficiency would need to be achieved, and the consequences of failing to do so.

- **What will be the role of the federal government, if any, in the NNMI after the end of the 10-year funding period?**
  As proposed, the NNMI would receive a one-time appropriation to be spent over 10 years. The Administration has not articulated its vision for the NNMI past the end of this period. Congress may wish to consider what role, if any, the federal government would play in coordinating or sustaining the network after the end of this period.
- **What role, if any, should the federal government play in advancing technologies that can contribute to manufacturing competitiveness?**
  Rapid technological advances may bring revolutionary changes to the manufacturing sector in the United States and abroad. Currently technologies such as grid computing, multi-core processors, massively parallel supercomputers, and new modeling software allow for more expansive, less expensive, and faster testing of designs. For example, automobile manufacturers are using these technologies to supplement physical crash testing of vehicles, thereby reducing costs, increasing passenger safety, and allowing for design considerations that might not otherwise have been possible.[49] According to Ford Motor Company:

> Prior to the first XJ prototype crashing into a barrier, Jaguar engineers performed more than 500 computer-simulated crash events using sophisticated crash-modeling software and this was followed up by physical tests.[50]

Further, new technologies, materials, processes, and design tools may allow for low-cost, high-customization, small lot-size production. In addition, new collaborative innovation models may open the possibility of making once-proprietary product design processes available to external creators. These new technologies and processes—and others that are likely to emerge from global research and development efforts—may displace existing industries, companies, and workers; shift value-added in the innovation process; and affect the competitive position of nations in manufacturing. In deciding whether to authorize and fund the NNMI, Congress may also wish to consider what the appropriate role of the federal government should be, if any, in advancing the U.S. position in manufacturing broadly (i.e., what boundaries should be set, if any, to define the appropriate roles of government and the private sector).

The Administration is currently seeking input from the stakeholder community with respect to a number of questions related to the foci, structure, governance, sustainability, education, and workforce development efforts of the NNMI and the individual IMIs. Congress may wish to explore these questions (see Appendix) and to hold hearings on what the Administration has learned from stakeholders in the course of its outreach activities.

APPENDIX. NIST *REQUEST FOR INFORMATION* QUESTIONS

The National Institute of Standards and Technology posed the following questions in a Request for Information published in the *Federal Register* (77 FR 26509) on May 4, 2012, seeking public input on the proposed National Network for Manufacturing Innovation. Commenters were advised that the questions were intended to assist in the formulation of comments, and were not to be construed as a limitation on the number of comments that interested persons may submit or as a limitation on the issues that may be addressed in such comments.

Technologies with Broad Impact

What criteria should be used to select technology focus areas?

What technology focus areas that meet these criteria would you be willing to co-invest in?

What measures could demonstrate that Institute technology activities assist U.S. manufacturing? What measures could assess the performance and impact of Institutes?

Institute Structure and Governance

What business models would be effective for the Institutes to manage business decisions?

What governance models would be effective for the Institutes to manage governance decisions? What membership and participation structure would be effective for the Institutes, such as financial and intellectual property obligations, access and licensing?

How should a network of Institutes optimally operate?
What measures could assess effectiveness of Network structure and governance?

Strategies for Sustainable Institute Operations

How should initial funding co-investments of the Federal government and others be organized by types and proportions?
What arrangements for co-investment proportions and types could help an Institute become self-sustaining?
What measures could assess progress of an Institute towards being self-sustaining? What actions or conditions could improve how Institute operations support domestic manufacturing facilities while maintaining consistency with our international obligations? How should Institutes engage other manufacturing related programs and networks? How should Institutes interact with state and local economic development authorities? What measures could assess Institute contributions to long term national security and competitiveness?

Education and Workforce Development

How could Institutes support advanced manufacturing workforce development at all educational levels?
How could Institutes ensure that advanced manufacturing workforce development activities address industry needs?
How could Institutes and the NNMI leverage and complement other education and workforce development programs?
What measures could assess Institute performance and impact on education and workforce development?
How might institutes integrate R&D activities and education to best prepare the current and future workforce?

End Notes

[1] National Association of Manufacturers website, *Facts About Manufacturing*, http://www.nam.org/Statistics-And-Data/Facts-About-Manufacturing/Landing.aspx
[2] For more information, see CRS Report R41898, *Job Creation in the Manufacturing Revival*, by Marc Levinson.
[3] Bureau of Labor Statistics, U.S. Department of Labor, Current Employment Statistics survey database, data for manufacturing employment, all employees, seasonally-adjusted, http://data.bls.gov/cgi-bin/dsrv.
[4] For more information, see CRS Report R41712, *"Hollowing Out" in U.S. Manufacturing: Analysis and Issues for Congress*, by Marc Levinson.
[5] See footnote 2.
[6] For additional information about the recession and the subsequent pace of economic growth, see CRS Report R41444, *Double-Dip Recession: Previous Experience and Current Prospect*, by Craig K. Elwell.
[7] National Science and Technology Council, Executive Office of the President, The White House, *A National Strategic Plan for Advanced Manufacturing*, February 2012, p. 1, http://www.whitehouse.gov/sites/default/files/microsites/ostp/ iam_advancedmanufacturing_strategicplan_2012.pdf.

[8] National Network for Manufacturing Innovation, http://www.manufacturing.gov/amp/ nnmi.html.

[9] Testimony of Patrick D. Gallagher, Under Secretary for Standards and Technology, U.S. Department of Commerce, before the U.S. Congress, House Committee on Science, Space, and Technology, Subcommittee on Technology and Innovation, *Assembling the Facts: Examining the Proposed National Network for Manufacturing Innovation*, 112th Cong., 2nd sess., May 31, 2012.

[10] National Institute of Standards and Technology, "Request for Information on Proposed New Program: National Network for Manufacturing Innovation (NNMI)," *Federal Register*, Vol. 77, No. 87, pp. 26509-26511, May 4, 2012, https://federalregister.gov/a/2012-10809.

[11] U.S. Department of Commerce, FY2013 Budget in Brief, February 2012, p. 123, http://www.osec.doc.gov/bmi/ budget/FY13BIB/fy2013bib_final.pdf.

[12] *Federal Register*, Vol. 77, No. 87, pp. 26509-26511, May 4, 2012, https://federalregister.gov/ a/2012-10809.

[13] In a telephone conversation on July 6, 2012, NIST indicated that the funding horizon may be reduced to seven years. In addition, the hearing charter for the House Committee on Science, Space, and Technology, Subcommittee on Technology and Innovation's hearing on *Assembling the Facts: Examining the Proposed National Network for Manufacturing Innovation*, May 31, 2012, stated that federal support for the institutes would span a period of five to seven years.

[14] Testimony of Patrick D. Gallagher, Under Secretary for Standards and Technology, U.S. Department of Commerce, before the U.S. Congress. See footnote 2.

[15] Office of Management and Budget, Executive Office of the President, *Fiscal Year 2013 Budget of the U.S. Government*, February 2012, Table S-9, p. 217.

[16] *Federal Register*, Vol. 77, No. 87, pp. 26509-26511, May 4, 2012, https://federalregister. gov/a/2012-10809.

[17] Testimony of Patrick D. Gallagher, Under Secretary for Standards and Technology, U.S. Department of Commerce, before the U.S. Congress. See footnote 2.

[18] The AMNPO is an interagency body hosted by the DOC's NIST, and includes DOD, DOE, NSF, the National Aeronautics and Space Administration, and other agencies. http://www.nist.gov/director/nnmi-050412.cfm

[19] *Federal Register*, Vol. 77, No. 87, pp. 26509-26511, May 4, 2012, https://federalregister.gov/ a/2012-10809.

[20] U.S. Congress, House Committee on Science, Space, and Technology, Subcommittee on Technology and Innovation, *Assembling the Facts: Examining the Proposed National Network for Manufacturing Innovation*, 112th Cong., 2nd sess., May 31, 2012.

[21] Council on Competitiveness, *Make: An American Manufacturing Movement*, Washington, DC, December 2011, pp. 63-64, http://www.compete.org/images/uploads/File/ PDF%20Files/ USMCI_Make.pdf. "

[22] The Information Technology and Innovation Foundation, *A Charter for Revitalizing American Manufacturing*, Washington, DC, 2011, p. 2; and "ITIF Welcomes President Obama's Proposal on Manufacturing Innovation," press release, March 9, 2012, http://www.itif.org/pressrelease/itif-welcomes-president-obamas-proposal-manufacturinginnovation.

[23] President's Council of Advisors on Science and Technology, *Report to the President on Ensuring American Leadership in Advanced Manufacturing*, Washington, DC, June 2011, http://www.whitehouse.gov/sites/ default/files/microsites/ostp/pcast-advanced-manufacturing -june2011.pdf.

[24] Ibid.

[25] President Barack Obama, *Remarks by the President on Manufacturing and the Economy*, The White House, Petersburg, VA, March 9, 2012, http://www.whitehouse.gov/the-press-office/2012/03/09/remarks-president-manufacturing-and-economy.

[26] *A Pilot Institute for the National Network for Manufacturing Innovation (NNMI)*, Solicitation Number BAA-12-17- PKM, Department of the Air Force, Air Force Materiel Command, May 8, 2012, https://www.fbo.gov/?s= opportunity&mode=form&id=2bbada5cac4ab 97438dc3f57fed050d0.

[27] Ibid.

[28] The BAA restricts the use of federal funding to direct support of the goals of the institute, for applied research, education and training, and infrastructure development, and explicitly prohibits the use of government funds to build buildings or to buy land or facilities.

[29] Ibid.

[30] Testimony of Patrick D. Gallagher, Under Secretary for Standards and Technology, U.S. Department of Commerce, before the U.S. Congress, May 31, 2012.

[31] According to the BAA, non-U.S. organizations may be allowed to use the Institute, subject to the sensitivity of the applied research.

[32] For additional information on the Small Business Innovation Research Program, see CRS Report 96-402, *Small Business Innovation Research (SBIR) Program*, by Wendy H. Schacht.

[33] For additional information on the Manufacturing Extension Partnership program, see CRS Report 97-104, *Manufacturing Extension Partnership Program: An Overview*, by Wendy H. Schacht.

[34] *A Pilot Institute for the National Network for Manufacturing Innovation (NNMI)*, Solicitation Number BAA-12-17- PKM, Department of the Air Force.

[35] Testimony of Patrick D. Gallagher, Under Secretary for Standards and Technology, U.S. Department of Commerce, before the U.S. Congress, May 31, 2012.

[36] *Federal Register,* Vol. 77, No. 87, pp. 26509-26511, May 4, 2012, https://federal register.gov/a/2012-10809.

[37] NIST website, http://www.nist.gov/director/nnmi-050412.cfm.

[38] *Designing for Impact I: Workshop on Building the National Network for Manufacturing Innovation: Summary of Results,* Energetics Incorporated, as prepared for the Advanced Manufacturing National Program Office, National Institute of Standards and Technology, June 12, 2012, http://manufacturing.gov/amp/rpi-summary-of-results.pdf.

[39] Testimony of Patrick D. Gallagher, Under Secretary for Standards and Technology, U.S. Department of Commerce, before the U.S. Congress, May 31, 2012.

[40] Mandatory spending is controlled by laws other than appropriations acts, often through authorizing legislation. Authorizing legislation establishes or continues the operation of a federal program or agency, either indefinitely or for a specified period. In contrast, discretionary spending is provided and controlled through the annual appropriations process. For additional information on mandatory funding, see CRS Report RL33074, *Mandatory Spending Since 1962*, by D. Andrew Austin and Mindy R. Levit.

[41] See footnote 40 for information on mandatory funding and how it differs from discretionary funding.

[42] For additional information see CRS Report 95-36, *The Advanced Technology Program*, by Wendy H. Schacht.

[43] For additional information, see CRS Report 97-104, *Manufacturing Extension Partnership Program: An Overview*, by Wendy H. Schacht.

[44] See footnote 19.

[45] Testimony of Patrick D. Gallagher, Under Secretary for Standards and Technology, U.S. Department of Commerce, before the U.S. Congress. See footnote 2.

[46] P.L. 100-418 uses the term "Regional Centers for the Transfer of Manufacturing Technology" to what is now called the Manufacturing Extension Partnership program.

[47] P.L. 100-418.

[48] P.L. 105-309. For further discussion, see CRS Report 97-104, *Manufacturing Extension Partnership Program: An Overview*, by Wendy H. Schacht.

[49] Deborah Wince-Smith, "High Performance Computing for All," *Issues in Science and Technology*, Summer 2009.

[50] Ford Motor Company website, http://media.ford.com/article_download.cfm?article_id=14028.

INDEX

A

B

D

E

F

G

H

I

J

K

L

M

Q

R

S

T